Topics in Control and its Applications

Springer-Verlag London Ltd.

Daniel E. Miller and Li Qiu (Eds.)

Topics in Control and its Applications

A Tribute to Edward J. Davison

 Springer

Daniel E. Miller, PhD
University of Waterloo, Department of Electrical and Computer Engineering,
Waterloo, Ontario, Canada. N2L 3G1

Li Qiu, PhD
Hong Kong University of Science and Technology, Department of Electrical and
Electronic Engineering, Clear Water Bay, Kowloon, Hong Kong

ISBN 978-1-4471-1158-0

British Library Cataloguing in Publication Data
Topics in control and its applications
    1. Control theory - Congresses
    I. Miller, Daniel E.   II. Qiu, Li
    629.8'312
ISBN 978-1-4471-1158-0

Library of Congress Cataloging-in-Publication Data
Topics in control and its applications : a tribute to Edward J.
    Davison / Daniel E. Miller and Li Qiu (eds.).
        p.   cm.
    Includes bibliographical references.
    ISBN 978-1-4471-1158-0   ISBN 978-1-4471-0543-5 (eBook)
    DOI 10.1007/978-1-4471-0543-5
    1. Automatic control.   I. Davison, Edward J., 1938-   .
II. Miller, Daniel E., 1962-   .   III. Qiu, Li, 1961-   .
TJ211.T67   1999   98-52347
629.8--dc21   CIP

© Springer-Verlag London 1999
Originally published by Springer-Verlag London Limited in 1999

Typesetting: Camera ready by editors

69/3830-543210 Printed on acid-free paper

Contents

Preface

Prof. Edward J. Davison (Ted) is a man of many talents. He started his career in music, veered into engineering physics for his undergraduate degree and moved to applied mathematics for his MASc degree; although accepted to do a PhD in cosmology, he ended up doing a PhD at Cambridge University in the area of systems control under the direction of the renowned Prof. Howard Rosenbrock. This breadth of ability underscores his ensuing thirty-five years of work in systems control, which spans the very applied to the highly theoretical. Equally important, he is arguably one of the most enthusiastic proponents of the area — he argued in the 1997 Bode Lecture that, unlike popular opinion, theory is lagging practise, with many exciting areas of science and engineering requiring input from the systems control area.

Ted's contributions to systems control are many. His first journal paper was on the area of model reduction, which became a "citation classic". He is commonly regarded to have introduced the word "robust" to the area, and has published extensively in the area of robust control, ranging from his 1976 paper on the robust servomechanism problem to the more recent paper on computing the real stability radius. He was in on the ground floor of the control of large scale systems, introducing such fundamental notions as decentralized fixed modes, and posing and solving the decentralized robust servomechanism problem. He has made contributions to determining fundamental benefits and limitations of adaptive control.

Throughout his career Ted has been acutely concerned with computations and numerical methods, having devised methods to solving "stiff" differential equations and devising computer aided optimal controller design routines. Besides his numerous consulting activities, he has done research on applications in a number of areas: in mechanical systems, such as the control of large flexible space structures; in chemical engineering systems, such as the control of distillation columns; in chemical systems, such as the modelling of polymeriza tion; in biological systems, such as developing models of cell behaviour, which was described in part in Time Magazine in 1974.

In addition to his academic contributions, Ted has put a lot of time and energy into the control systems community. His many professional activities include that of vice-president of the the IEEE Control Systems Society (1979 – 1981), president of the IEEE Control Systems Society (1983), and chairman of the IFAC Theory Committee (1988 – 1990).

As former graduate students of Ted, we benefited directly from his guidance. The whole area of control systems has benefited from his scholarly contributions, the leadership he provided in various professional capacities, and his infectious enthusiasm for our area of research as well as life in general.

This year Ted turned 60. In honor of this event we hosted a workshop at the Fields Institute of the University of Toronto during June 29-30, 1998.

A number of papers were presented during the workshop, and most are in some way related to Ted's work. The topics fall into four broad classes: robust control, decentralized control, control applications, and various topics in the theory of signals and systems.

We would like to thank the authors for their enthusiastic participation. We are grateful to Dr. Don Dawson of the Fields Institute for hosting the workshop and providing support staff, and to Prof. David Burns of the University of Waterloo and Prof. Safwat Zaky of the University of Toronto for providing financial support. We would like to thank Brenda Law of the Fields Institute who was instrumental in the smooth functioning of the workshop, and Prof. Diane Kennedy of Ryerson Polytechnic University for helping to arrange accommodations and the banquet. we are grateful for the support of Springer-Verlag, who is publishing the conference proceedings to ensure wide dissemination; we are especially grateful for the support of Nicholas Pinfield, Engineering Editor, Springer-Verlag London, and for the editorial help from Anne Neill and Elaine Fuoco. Finally, we would like thank Lau Mei Sum of Hong Kong University of Science and Technology for typesetting part of the book.

Waterloo, Hong Kong   *Daniel E. Miller*
November 1998   *Li Qiu*

Biography of Edward J. Davison

Edward J. Davison was born in Toronto, Canada in 1938. He received the A.R.C.T. degree in piano from the Royal Conservatory of Music of Toronto in 1958, and the B.A.Sc. degree in Engineering-Physics and the M.A. degree in Applied Mathematics from the University of Toronto in 1960 and 1961 respectively. In 1964 he received the Ph.D. degree, and in 1977 the Sc.D. degree from Cambridge University, England.

From 1964-1966 he was with the University of Toronto; in 1966-67 he was with the University of California, Berkeley, in the Department of Electrical Engineering and Computer Science, and since then, he has been associated with the Department of Electrical and Computer Engineering, University of Toronto. His research interests include the study of linear systems theory, large scale systems, CAD of control systems, computational methods, adaptive control, intelligent control, industrial process control, large flexible space structures and biological system theory.

Dr. Davison was Associate Editor from 1974 to 1976, Guest Associate Editor in 1977-1978, 1982-1983, and Consulting Editor in 1985, of the *IEEE Trans. on Automatic Control.* He has been an Associate Editor of *Automatica* from 1974-1987, and of *Large Scale Systems: Theory and Applications* from 1979-1990, and has been a member of the Editorial Board of *Optimal Control Applications and Methods* since 1983. He was Vice-President (Technical Affairs) in 1979-1981, President-Elect in 1982, and President in 1983, of the IEEE Control Systems Society. He was Vice-Chairman of the International Federation of Automatic Control (IFAC) Theory Committee in 1978-1987, Chairman of the IFAC Theory Committee in 1988-1990, Vice-Chairman of the IFAC Technical Board in 1990-1993, a member of the IFAC Council in 1991-1996, and at present has been serving as Vice-Chairman of the IFAC Policy Committee since 1996.

Dr. Davison has received several awards, including the Athlone Fellowship (Cambridge University) 1961-1963, the National Research Council of Canada's E.W.R. Steacie Memorial Fellowship 1974-1977, the Canada Killam Research Fellowship 1979-1980, 1981-1983, two IEEE Transactions on Automatic Control Outstanding Paper Awards, and a Current Contents Classic Paper Citation. He was elected a Fellow of the Institute of Electrical and Electronic Engineers in 1977, a Fellow of the Royal Society of Canada in 1977, and an Honorary Professor of Beijing Institute of Aeronautics and Astronautics in 1986. In 1984, he received the IEEE Centennial Medal and was elected a Distinguished Member of the IEEE Control Systems Society. In 1996 he received the Outstanding Member Service Award from IFAC, and in 1998 he was elected a Member of the Academy of Nonlinear Sciences in Moscow, Russia.

Dr. Davison is a designated Consulting Engineer of the Province of Ontario since 1979, and currently is President of the consulting firm Electrical Engineering Consociates (EEC) situated in Toronto, Canada.

In 1993, he was awarded the triennial Quazza Medal from the International Federation of Automatic Control (IFAC), and in 1997 he was awarded the IEEE Control Systems Society's Hendrik W. Bode Lecture Prize.

Robust Control for Car Steering

Jürgen Ackermann

DLR, German Aerospace Center, Institute of Robotics and System Dynamics,
Oberpfaffenhofen, D-82234 Wessling, Germany
Email: Juergen.Ackermann@dlr.de

Abstract. The human driver is very good at controlling vehicle dynamics, if the decisions can wait for a second. If, however, the dynamics are very fast, like in the beginning of skidding or rollover, then an automatic driver support system reacts faster and more precisely and reliably than the driver with his reaction time and overreaction. Such a control system should not be turned on only after a safety critical situation has been detected, which also takes some time. The control system should be in continuous operation for immediate reaction to a disturbance, then it also provides improved comfort under continuous disturbances, e.g. from a trailer or gusty wind.

The presented concept is based on feedback of the measured yaw rate to an actuator that mechanically adds a small corrective steering angle to the steering angle commanded by the driver. The control algorithm is derived from the idea that the driver commands a lateral acceleration to a point mass in order to keep it on the planned path. This motion implies a reference yaw rate. Deviations of the measured yaw rate from the reference yaw rate are fed back to control the corrective steering angle. A special feature of the presented control law is its robustness with respect to unknown lateral tire forces, and uncertain velocity and mass. The result is a robust separation of the lane keeping task of the driver from the yaw stabilization task of the automatic control system. These two separated subsystems can be further improved independently by conventional control designs.

The safety improvements have been demonstrated by μ-split braking and sidewind tests with a steer-by-wire car.

1 Historic Background

Active steering systems for automobiles have been studied for a long time now. 30 years ago, Kasselmann and Keranen [1] designed an active control system that measures the yaw rate by a gyro and uses proportional feedback to generate an additive steering input for the front wheels, see Fig. 3. Test results indicated "that the system greatly reduces the lateral motion of a vehicle subjected to wind gusts or rutted-road driving. It also reduces driver-steering inputs in number and magnitude, not only under these special conditions but also in normal expressway driving". Regarding the use of only proportional feedback the following simulation experience is reported: "The use of dynamic compensation was evaluated and was ultimately rejected as adding needless complexity to a satisfactorily performing circuit." The proportional gain was scheduled by the velocity such that it increases up to a velocity of 30 miles per hour. For the position of the actuator for the additive steering angle the following design decisions were made: "The location integral with the power-steering gear was rejected because it violated the concept of making adaptive steering available as add-on accessory. The steering-column location was attractive because it permitted the use of a low-powered actuator, it was ultimately rejected, however, because of the difficulty of retaining limited-authority operation and also because of the poor dynamic characteristics of the power-steering gear. The linkage location was accepted because good actuator dynamic response could be achieved without upsetting the driver. The poor dynamic characteristics and mechanical advantage of the power-steering gear also served to isolate the driver from actuator reaction forces." A hydraulic actuator for translational shift in the steering linkage was chosen to generate the additive steering angle, which is limited to ±3 degrees of front-wheel movement. This early Bendix study never made it to a product. Some of the above ideas are, however, still relevant for future active front-wheel steering systems.

In the early 80's studies on automatic track following for busses were initiated in Germany, a good example was published by Darenberg [2]. The author [3] contributed a robust steering control law, where robustness refers to variations of operating conditions (road and tire contact, mass and velocity of the vehicle). In [4] it was shown that the robustness and tracking accuracy can be drastically improved by additional feedback of the yaw rate to the steering actuator.

During the 80's and early 90's four-wheel steering became a hot topic, see for example the survey by Furukawa et al. [5]. Typically the front wheel steering was unchanged and an hydraulic or electric actuator for additional rear-wheel steering was used. Initially only feedforward control laws (some with gain scheduling) have been employed; later also feedback from vehicle dynamics sensors was included in order to reduce the effect of disturbance torques and parameter uncertainty, an example is the work by Hirano and Fukatani at Toyota [6].

In 1990 the author [7,8] proposed a concept for feedback of the yaw rate to active front and rear wheel steering. The first design goal is a clear separation of the track following task of the driver from the automatic yaw stabilization that balances disturbance torques around a vertical axis. This goal is achieved in a robust way, i.e. independent of the operating conditions. The robust decoupling effect is obtained by integral feedback of the yaw rate to front-wheel steering. The undesired side effect of reduced yaw damping at high velocity was removed by yaw rate feedback to rear-wheel steering. Both the front-wheel and the rear-wheel control may be improved independently of each other by conventional control engineering methods. For example accelerometer feedback to the front axle can speed up the steering response, while velocity scheduled yaw rate feedback to rear-wheel steering allows a specification on the desired velocity dependence of yaw damping [9].

The main obstacle for an implementation was the cost of the hardware. Meanwhile the cost of the yaw rate sensors has come down because they are also used in individual wheel braking systems like ESP. The cost of rear-wheel steering can be avoided by alternative ways of improving yaw damping via front-wheel steering. What remains is the actuator for mechanical addition of the feedback-controlled front wheel steering angle. A planetary gear for insertion into the steering column is offered by Bosch [10]. The problems associated with this actuator location have already been discussed in the Bendix paper. Constructive proposals for a translational shift in the steering linkage have been made by Fleck [11] for a hydraulic actuator and by the author and colleagues [12] for an electric actuator. The latter actuator employs a spindle drive with extremely low friction, which was developed at DLR for robotic applications.

The disturbance rejection properties of the robust steer control have been experimentally verified on a BMW test car equipped with steer-by-wire [13]. In the meantime several improvements of the robust control algorithm have been made, that will be reported in the present paper. An important conceptual improvement is the use of a "fading integrator" instead of the ideal integrator. Thereby the driver support system has its full effect immediately after the occurrence of a disturbance torque, thereafter the control of longer lasting disturbances is softly transferred to the driver, such that the steady-state behavior of the car is unchanged by the control system.

2   Separation of Path Following and Yaw Stabilization

Steering a car involves two tasks, a primary task of path following and a secondary task of yaw stabilization under yaw disturbance torques. For path following the driver keeps the car – considered as a single point mass m_P – on top of his planned path, in other words: he applies a lateral acceleration a_{yP} to the mass m_P in order to reorient the velocity vector v such that it remains tangential to his planned path, see Fig. 1. The second task results from the fact, that the real car is not a point mass; it may be described as a body with moment of inertia J, in Fig. 1 represented by a second rigidly connected mass m_R. The yaw rate r of the car is not only excited by a_{yP} in a way that the driver is used to, but also by disturbance torques M_{zD} resulting for example from crosswind or from a flat tire or from braking on ice, where we have asymmetric friction coefficients at the left and right tires.

The driver has to compensate this disturbance torque by counteracting at the steering wheel in order to provide disturbance attenuation. This is the more difficult task for the driver because the disturbance input M_{zD} comes as a surprise to him; that means it takes him about a second of reaction time to recognize the situation and to decide what to do, and then he may even overreact and make things worse.

We want to leave the task of path-following with the driver, the second task of disturbance attenuation will be assigned to an automatic control system. First of all, we have to decouple the secondary yaw dynamics such that they do not influence the primary path-following dynamics. The automatic control system for the yaw rate r should not interfere with the path-following task of the driver.

What we want to achieve in system theoretic terms is to make r unobservable from the lateral acceleration a_{yP}. This decoupling is unilateral; there will be, of course, an influence of a_{yP} on r, otherwise the driver could not command the car to enter a curve. But the driver commands a desired yaw rate only indirectly via a_{yP}, he should be concerned directly only with a_{yP}.

Decoupling has been known since decades, but the classic results apply to nominal plant parameters and a given output. The problem is, however, that we want to achieve robust decoupling. It should in particular be robust with respect to the road surface condition and car velocity.

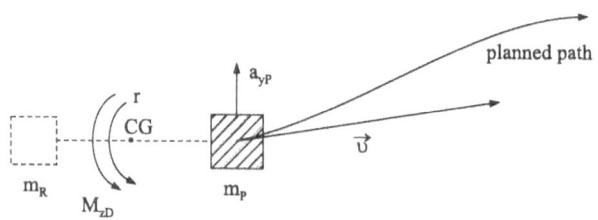

Fig. 1. Path following and yaw stabilization.

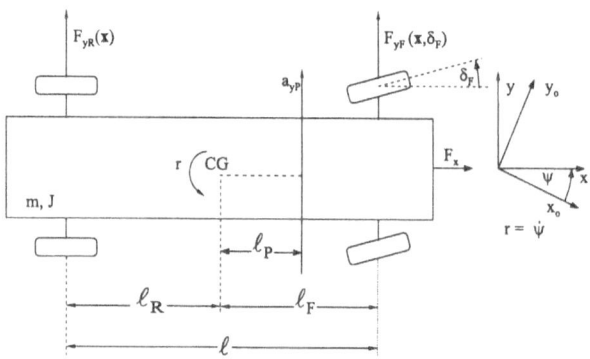

Fig. 2. Unknown rear and front axle lateral forces act on the car body with mass m and moment of inertia J.

Road surface conditions can change very rapidly, for example in springtime when most of the snow has melted but there are some remaining ice patches on the otherwise dry road. There is no time for identification of the friction coefficient μ and for adaptation of the controller, thus the controller has to be robust with respect to u. The velocity is not changing so quickly, also a velocity signal is available from the ABS brakes or a speedometer and can be used for scheduling the feedback controller.

The car is modelled as a rigid body with mass m and moment of inertia J with respect to a vertical axis through the center of gravity CG, see Fig. 2. The chassis coordinate system x, y is rotated by the yaw angle ψ with respect to an inertially fixed coordinate system x_0, y_0. The yaw rate is $r = \dot{\psi}$, it is measured by a yaw rate sensor. The yaw rate will be used as one of the state variables in the state vector $\mathbf{x}$.

The most important uncertainties in modelling the motion of the vehicle are the lateral forces F_{yR} at the rear axle and F_{yF} at the front axle. Both of them depend on the state $\mathbf{x}$ of the car. The front lateral force also depends on the front wheel steering angle δ_F. There are tire models in the automotive literature that give the lateral forces. However, they describe the forces in terms of other quantities which we do not know as well, for example the road friction coefficient μ. Therefore, we consider the forces F_{yR} and F_{yF} directly as the unknown quantities.

Now remember that we want to remove the influence of r on the lateral acceleration. If we choose, for example, the lateral acceleration at the CG as the output to be decoupled, then we have no chance to achieve this goal in view of the unknown forces $F_{yR}(\mathbf{x})$ and $F_{yF}(\mathbf{x}, \delta_F)$. The first key idea is to choose a position at a distance ℓ_P in front of the CG such that the lateral acceleration a_{yP} at this point does not depend on $F_{yR}(\mathbf{x})$. Thus, unlike in conventional decoupling, we also have to choose the output position where robust unilateral decoupling is possible. It is a calculation of a few lines [14] to find the position

$$\ell_P = \frac{J}{m\ell_R}. \tag{1}$$

At this point the lateral acceleration due to tire forces is

$$a_{yP} = \frac{\ell}{m\ell_R} F_{yF}(\mathbf{x}, \delta_F) \tag{2}$$

where $\ell = \ell_R + \ell_F$ is the wheelbase, see Fig. 2.

The second key idea is to compensate the influence of r (which is part of $\mathbf{x}$) by δ_F in the argument of $F_{yF}(\mathbf{x}, \delta_F)$. If this argument is independent of r, then also the unknown force F_{yF} and therefore a_{yP} is independent of r.

The plant input is the front wheel steering angle δ_F. In the active steering system of Fig. 3 it is composed of a conventional steering angle δ_S commanded by the driver and an additive steering angle δ_C generated by the feedback controller, such that

$$\delta_F = \delta_S + \delta_C. \tag{3}$$

For robust unilateral decoupling δ_C must compensate the influence of r on F_{yF}. Both δ_C and r enter into F_{yF} via the tire slip angle α_F which is illustrated in Fig. 4. (Note that the two front wheels are combined into one wheel in the center of the axle. This is known as the single-track model.)

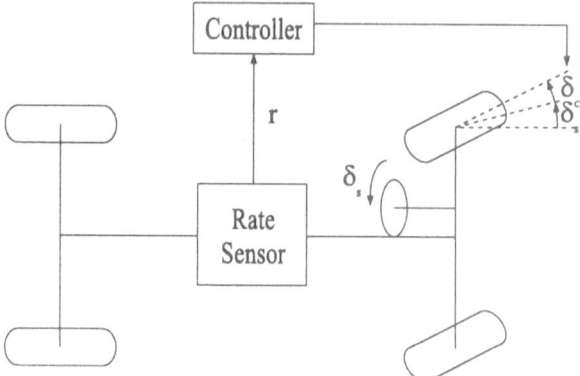

Fig. 3. The steering angle $\delta_F = \delta_S + \delta_C$ is composed of the command δ_S from the driver and the feedback controlled additional angle δ_C.

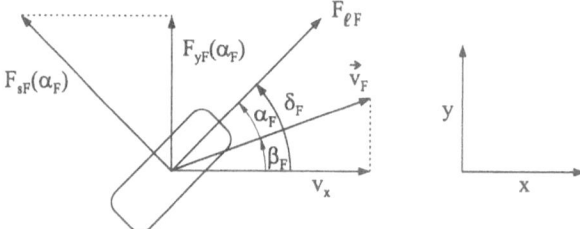

Fig. 4. Angles and forces at the front wheel of a single-track model.

The longitudinal tire force $F_{\ell F}$ is commanded by brake and throttle and does not depend on r. The side force F_{sF} is a function of the tire slip angle α_F between the tire direction and the local velocity vector v_F.

Here we make the simplifying assumption that the steering angle δ_F is small and therefore approximately $F_{yF}(\alpha_F) = F_{sF}(\alpha_F)$, and equation (2) becomes

$$a_{yP}(\alpha_F) = \frac{\ell}{m\ell_R} F_{sF}(\alpha_F). \tag{4}$$

Thus the yaw rate r does not influence a_{yP} if and only if r does not influence α_F. We want to make the yaw rate unobservable from the front tire slip angle.

Another angle of interest here is the chassis slip angle β_F at the front axle, see Fig. 4. It may be used as another state variable such that we have the state vector

$$x = \begin{bmatrix} v_x \\ \beta_F \\ r \end{bmatrix} \tag{5}$$

where $v_x = |v_F| \cos \beta_F$ is the velocity component in x-direction, which can be measured by the rear tire ABS sensors. The state equations of the single-track

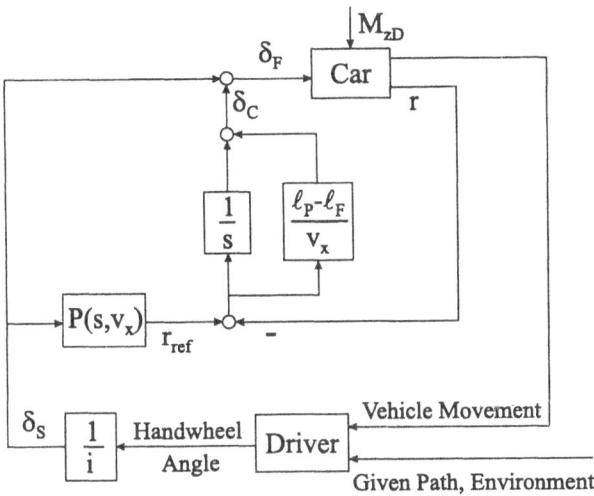

Fig. 5. Block diagram of Robust Steer Control.

model then have the form

$$\dot{\mathbf{x}} = f(\mathbf{x}, \delta_F) \qquad (6)$$

and we are interested in the output

$$\alpha_F = \delta_F - \beta_F = \delta_S + \delta_C - \beta_F. \qquad (7)$$

We cannot measure β_F at reasonable cost, otherwise $\delta_C = \beta_F$ would be a feasible decoupling control law. From the differential equation (6), however, we know $\dot{\beta}_F$. The third key idea is to use a backstepping approach and to do the compensation at the input of an integrator with input $\dot{\delta}_C$. The details of the derivations are given in [14]. The Laplace form of a simplified controller is

$$\delta_C(s) = \left(\frac{1}{s} + \frac{\ell_P - \ell_F}{v_x} \right) [P(s, v_x)\delta_S(s) - r(s)]. \qquad (8)$$

A block diagram of the overall steering system is shown in Fig. 5.

The steering gear ratio is i. The forward velocity v_x is used for scheduling the proportional part of the controller $(\ell_P - \ell_F)/v_x$ and the prefilter $P(s, v_x)$. The proportional part (or the entire feedback path) is turned on softly when a velocity $v_{x\min}$ is reached such that division by zero is avoided.

It is shown in [14] that the differential equation for the front tire slip angle becomes

$$\dot{\alpha}_F = - F_{sF}(\alpha_F) \frac{\ell}{m\ell_R v_x} + r_{\text{ref}} + \frac{\ell_P - \ell_F}{v_x} \dot{r}_{\text{ref}} + \dot{\delta}_S \qquad (9)$$

where

$$r_{\text{ref}}(s) = P(s, v_x)\delta_S(s).$$

Thus α_F depends only on the external input δ_S from the steering wheel, but not on r, the yaw rate r has become unobservable from the front tire slip angle α_F and therefore also from the lateral acceleration a_{yP}. Thus, we have achieved robust unilateral decoupling.

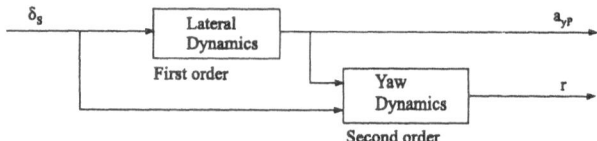

Fig. 6. The yaw rate r is unobservable from the lateral acceleration a_{yP}.

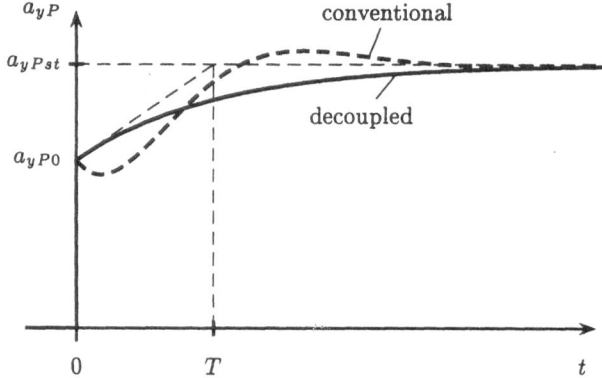

Fig. 7. Open- and closed-loop responses to a steering wheel step.

The effect of the decoupling controller is illustrated by the signal flow diagram of Fig. 6. The driver is only concerned with the first order subsystem "lateral dynamics". Through this system he commands the lateral acceleration a_{yP}. In the conventional car the input-output relationship from δ_S to a_{yP} is of second order. Figure 7 shows a comparison of linearized a_{yP}-responses to a steering wheel step input. The conventional vehicle has an oscillatory second order response that depends heavily on the uncertain parameters. The decoupled vehicle also has relative degree zero and the same immediate reaction of the lateral acceleration to a steering wheel step. Then follows an exponential, first order transient with small dependency of the time constant T on uncertain parameters. The identical immediate reaction a_{yP0} is guaranteed by the direct throughput from δ_S to δ_F. The identical steady-state value a_{yPst} was tuned here by the prefilter P, see Fig. 5. This point will be discussed later.

The unilateral decoupling property is robust with respect to the road surface, unknown nonlinear tire characteristics, driving speed v_x (assumed con-

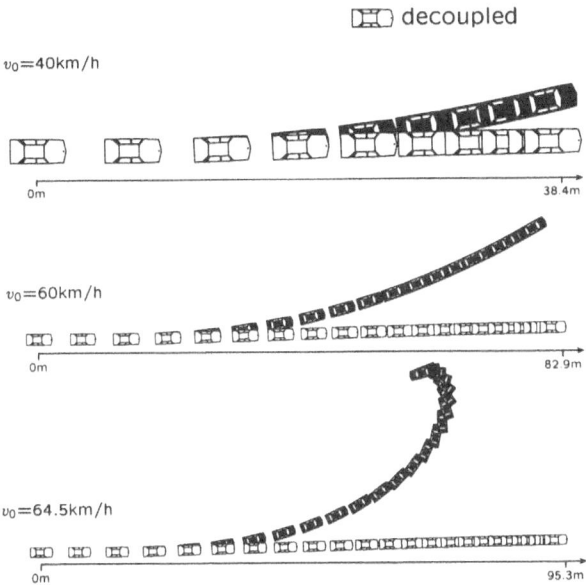

conventional
decoupled

$v_0=40$km/h

0m   38.4m

$v_0=60$km/h

0m   82.9m

$v_0=64.5$km/h

0m   95.3m

Fig. 8. μ-split-braking at various initial velocities ($\mu_{left} = 0.8, \mu_{right} = 0.1$)

stant or slowly varying) and vehicle mass provided the position $\ell_P = J/m\ell_R$ is unchanged (or tuned after each start of the car).

3   Nonlinear Simulations and Road Tests on Disturbance Attenuation

The disturbance attenuation property of the decoupling control concept is a consequence of the fact that yaw disturbance torques M_{zD} directly enter only into $\dot{r}$. The integrated effect of M_{zD} on r is removed from a_{yP} by making r unobservable from a_{yP}. In this section the disturbance attenuation effect is shown quantitatively by nonlinear simulations and road tests.

In the stroboscopic view seen in Fig. 8 we compare the conventional un-controlled vehicle and the vehicle with decoupling controller. The maneuver is μ-split braking with friction coefficients $\mu = 0.8$ at the left tires and $\mu = 0.1$ at the right tires. The driver pushes the brake and keeps the steering wheel straight, $\delta_S \equiv 0$. The asymmetry of braking forces causes a disturbance torque which rotates the car. At an initial velocity of 40 km/h the conventional car moves to the left, the controlled car is initially also rotated to the left but is then rotated back to the initial yaw angle. There remains a small parallel displacement that is easily correctable by the driver. At an initial velocity of 60 km/h we get more of the same, the conventional car has a larger deviation now. In the third simulation for 64.65 km/h we see an interesting nonlinear effect that results from tire force saturation. We do not get more of the same, the behavior of the car changes drastically and that is a bad surprise for the driver. In contrast, the controlled car keeps a good safety margin from tire saturation.

The disturbance attenuation effects have also been experimentally verified. Fig. 9 shows experimental results that we have obtained on the BMW test track near Munich [13]. Here the μ-split braking test is performed on a lane with different friction coefficients. The track is divided, in the left side of the picture there are water-flooded tiles with a friction coefficient $\mu = 0.1$, on the right side of the picture there is wet asphalt with a friction coefficient $\mu = 0.8$. The driver brakes and keeps the steering wheel straight as indicated by the white arrow at the steering wheel. From the video tape we have taken four snapshots each at 1 second intervals. The conventional car on the left series starts rotating in the second shot. Here the driver could notice that braking is irregular, but under normal driving conditions it would probably take the driver about one second of reaction time to decide what to do. After this second he is already on the asphalt with all four wheels in this test, so he does not really have a chance to correct the skidding motion. The right series shows our decoupled car. Initially the same happens as in the conventional car. In the second picture the car has a yaw rate that is measured by a rate sensor and fed back into the active steering system. During the next crucial second the car is brought back to its initial yaw angle. There only remains a small lateral displacement that is easily correctable by the driver a few seconds later.

Similar results have been obtained with crosswind experiments see Fig. 10. These experiments show the significant safety advantage of the decoupling control system.

4   Handling Improvement

We also did some hard handling maneuvers with the test car and they were not satisfactory. One of the reasons was the insufficient bandwidth of the steering servo motor and its rate limitations, but there are also control theoretic reasons that will be discussed in this section together with empirical improvements of the control law. Of course, we want to preserve the nice safety features under all modifications.

It was shown in [9] that the robust unilateral decoupling control system of Fig. 5 reduces the yaw damping at high velocities. For cars with additional rear-wheel steering this is no problem, because any desired dependency of the yaw damping on the velocity can be obtained by feedback of the yaw rate to rear wheel steering. The resulting additional forces at the rear axle have no influence on the lateral acceleration a_{yP} at the decoupling point, see Fig. 2. For cars without rear-wheel steering the yaw damping may be improved via active front-wheel steering. This, however, requires some compromise with the ideal robust decoupling concept.

Fig. 9. μ-split-braking at $v = 80\,\text{km/h}$ for conventional car (left) and for robustly decoupled car (right)

A promising approach is the replacement of the integral controller action (transfer function 1/s) by a "fading integrator" with transfer function

$$G_{fi}(s) = \frac{s}{s^2 + 2D_{fi}w_{fi}s + w_{fi}^2}. \tag{10}$$

The fading integrator has the same initial step response as the integrator, but the response later fades out to zero, see Fig. 11.

Empirically the parameters $w_{fi} = 1\,\text{rad/s}, D_{fi} = 1.5$ were chosen for Fig. 11. The decoupling and disturbance rejection properties of the ideal integrator are approximately preserved for the first 0.5 seconds, during which the driver cannot react anyway to unexpected disturbances.

Fig. 10. Crosswind acting on conventional car (left) and on robustly decoupled car (right).

The fading action is applied also to the proportional feedback path. The fading controller transfer function now becomes

$$G_{fc}(s) = \frac{(1 + \frac{\ell_P - \ell_F}{v_x} s)s}{s^2 + 2D_{fi}w_{fi}s + w_{fi}}. \tag{11}$$

The s-term in the numerator guarantees, that the controller output δ_C goes to zero in the response to a steering step (δ_S) or disturbance step M_{zD}. The advantages of the controller (11) compared with the controller of Fig. 5 are

i) The steady-state steering angle, e.g. in cornering or continuous crosswind is only determined by the direct throughput from δ_S to δ_F, i.e. it is identical to the steady-state steering angle of the conventional car.

Step responses

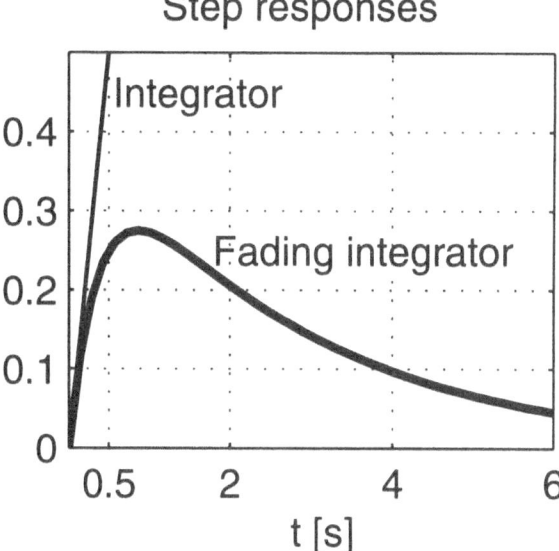

Fig. 11. Step responses of $1/s$ and $G_{fi}(s)$ (fading integrator with $w_{fi} = 1\,\mathrm{rad/s}$, $D_{fi} = 1.5$).

ii) The danger of actuator saturation is much reduced.
iii) Stability is determined by the long term behavior, and this is identical to that of the conventional car.

The response of the controlled car to a disturbance step M_{zD} is now such that the driver wins time. It suffices for disturbance torque compensation that the driver reacts about a second later and the effect of M_{zD} comes more softly. The immediate reaction to the disturbance is, however, close to the immediate reaction as seen in the experiments, Figs. 9 and 10. The response of the controlled car to a steering step input is such that both the immediate response and the steady-state response are as in the conventional car. The transients in between can be influenced by the prefilter $P(s, v_x)$. A favorable choice is

$$P(s, v_x) = \frac{K_P(v_x)}{1 + 0.1s} \tag{12}$$

where

$$K_P(v_x) = \frac{v_x}{\ell[1 + (v_x/v_{CH})^2]} \tag{13}$$

with the nominal characteristic velocity v_{CH} of the vehicle. The gain in chosen with the same velocity dependence as in the conventional car. The delay yields the same relative degree 1 in the transfer functions from δ_S to both r and r_{ref}, see Fig. 5. Thereby the step response in $r_{ref} - r$ starts with a ramp rather than a step. The prefilter (12),(13) results in a soft excitation of the yaw oscillation and thereby further reduces the effect of weak damping at high velocities. Further it reduces the requirements on the steering actuator for generating δ_C and adding it mechanically to δ_S. A more detailed analysis of the fading controller and the prefilter is given in [15].

An analysis of the linearized system in terms of step responses and Bode plots for the inputs δ_S and M_{zD} and the outputs r and a_{yP} can be found in [16].

An alternative approach for providing better yaw damping at high velocity is discussed in [14]. Beyond a critical velocity a gain scheduled additional feedback path is activated that preserves the factorization that was illustrated in Fig. 6. Only the damping of the yaw subsystem is increased, while the lateral dynamics remain almost unchanged.

5   Conclusions

In conclusion, we have developed the concept of robust unilateral decoupling of car steering dynamics. Its effect is that the driver has to care much less about disturbance attenuation. The important quick reaction to disturbance torques is done by the automatic feedback system. The yaw dynamics no longer interferes with the path following task of the driver. The safety advantages have been demonstrated in experiments with a test vehicle. By empirical improvements we have then modified the controller such that it preserves the robust decoupling advantages for the first 0.5 seconds after a disturbance and then returns the steering authority gradually back to the driver.

References

1. J. Kasselmann and T. Keranen, "Adaptive steering," *Bendix Technical Journal*, vol. 2, pp. 26–35, 1969.
2. W. Darenberg, "Automatische Spurführung von Kraftfahrzeugen," *Automobil-Industrie*, pp. 155–159, 1987.
3. J. Ackermann and S. Türk, "A common controller for a family of plant models," in *Proc. 21st IEEE Conf. Decision and Control*, (Orlando), pp. 240–244, 1982.
4. J. Ackermann and W. Sienel, "Robust control for automatic steering," in *Proc. American Control Conference*, (San Diego), pp. 795–800, 1990.
5. Y. Furukawa, N. Yuhara, S. Sano, H. Takeda, and Y. Matsushita, "A review of four–wheel steering studies from the viewpoint of vehicle dynamics and control," *Vehicle System Dynamics*, vol. 18, pp. 151–186, 1989.
6. Y. Hirano and K. Fukatani, "Development of robust active rear steering control," in *Proc. International Symposium on Advanced Vehicle Control*, pp. 359–375, June 1996.
7. J. Ackermann, "Verfahren zum Lenken von Straßenfahrzeugen mit Vorder- und Hinterradlenkung." Patent No. P 4028 320 Deutsches Patentamt München, Anmeldung 6.9.90, erteilt 18.02.93; European Patent 0474 130, US Patent 5375057, 1993.

8. J. Ackermann, "Robust car steering by yaw rate control," in *Proc. IEEE Conf. Decision and Control*, (Honolulu), pp. 2033–2034, 1990.

9. J. Ackermann, A. Bartlett, D. Kaesbauer, W. Sienel, and R. Steinhauser, *Robust control: Systems with uncertain physical parameters*. London: Springer, 1993.

10. W. Krämer and M. Hackl, "Potential functions and benefits of electronic steering assistance," in *XXVI Fisista Congress*, (Praha), June 17-21 1996.

11. R. Fleck, "Konstruktion einer aktiven Lenkung." Diplomarbeit, Technische Universität München, Lehrstuhl für Verbrennungskraftmaschinen und Kraftfahrzeuge, 1996. Durchgeführt bei BMW Technik GmbH, München.

12. J. Ackermann, T. Bünte, J. Dietrich, B. Gombert, and B. Willberg, "Aktuator zum Korrigieren eines über das Lenkrad eines Fahrzeugs an die Räder einer gelenkten Achse eingegebenen Lenkwinkels." Patent Application 197 50 585.6 Deutsches Patentamt München, filed Nov. 17, 1997.

13. J. Ackermann, T. Bünte, W. Sienel, H. Jeebe, and K. Naab, "Driving safety by robust steering control," in *Proc. Int. Symposium on Advanced Vehicle Control*, (Aachen, Germany), June 1996.

14. J. Ackermann, "Robust unilateral decoupling of the influence of the yaw motion on the lateral acceleration of a car," in *Proc. International Conference on Advances in Vehicle Control and Safety*, (Amiens, France), July 1998.

15. J. Ackermann and T. Bünte, "Handling improvement of robust car steering," in *Proc. International Conference on Advances in Vehicle Control and Safety*, (Amiens, France), July 1998.

16. J. Ackermann, "Robust control prevents car skidding," (Bode Prize Lecture 1996), *IEEE Control Systems Magazine*, vol. 17, pp. 23–31, June 1997.

Radial and Uniform Distributions in Vector and Matrix Spaces for Probabilistic Robustness

Giuseppe Calafiore[1], Fabrizio Dabbene[2], and Roberto Tempo[2]

[1] Dipartimento di Automatica e Informatica, Politecnico di Torino, Italy
[2] CENS-CNR, Politecnico di Torino, Italy
   Email: calafiore, dabbene, tempo@polito.it

Abstract. A number of recent papers focused on probabilistic robustness analysis and design of control systems subject to bounded uncertainty. This approach is based on randomized algorithms that require generation of vector and matrix samples. In this paper, we provide theoretical results as well as efficient algorithms for the generation of radial and uniform samples in norm-bounded sets.

1   Preliminaries

Recently, the robust control community studied several problems related to the so-called *probabilistic approach* for robustness analysis of uncertain control systems. For parametric uncertainty it can be easily shown that the number N of randomly generated samples required to meet specified accuracy and confidence is independent of the number of real parameters entering into the control system. This fact is an immediate consequence of the Law of Large Numbers and it is often used in Monte Carlo simulation; see [12,15]. More generally, if the control system is subject to structured and unstructured uncertainty, real or complex, the same conclusion holds. Therefore, in the standard M-Δ configuration [19], we can say that N is independent of the number and size of the blocks δ_i and Δ_k. Furthermore, bounds on the sample size can be easily computed [10,16] for probability estimation and worst-case performance problems. A subsequent line of research, aiming towards probabilistic design, utilizes concepts from Learning Theory [17]. A crucial advantage of this setting is the fact that the *problem structure* is used. That is, on the contrary of the Monte Carlo approach where the sample size depends only on the accurary and confidence required, in this framework the sample size is a function of other quantities such as, for example, the Mc Millan degree of the plant and compensator. More precisely, the problem structure is taken into account through a parameter called the VC-dimension; see [18] for further details.

   In both Monte Carlo and Learning Theory approaches, however, a crucial problem is to find efficient algorithms for sample generation in various sets. That is, given a certain multivariate probability density function f with compact support S, a key issue is the generation of N samples within S distributed according to f. In this paper, we provide specialized algorithms for the case when S is a norm-bounded set defined in vector and matrix spaces. This case is of interest in robustness analysis and has not been explicitly treated in the existing Monte Carlo literature, see e.g. [7,13] and references therein. As a starting point, the distribution f is assumed to be uniform over S. This assumption is supported by the fact that the uniform distribution coincides with the worst-case distribution in a certain class [2,3]. Additional properties of the uniform distribution have been shown in [1]. Finally, we show how to obtain generic radially symmetric distributions by an appropriate rescaling of the uniform samples.

We now describe more precisely the problems studied in this paper. Given a real or complex ball of radius $r > 0$

$$B(r) = \{x \in \mathbb{F}^n : \|x\|_p \leq r\}$$

where $\|\cdot\|_p$ is the standard ℓ_p norm, the objective is to generate vector samples uniformly in $B(r)$. Subsequently, we study extensions to uniform generation of matrix samples.

In Sect. 2, we introduce some preliminary concepts about probability densities and random sample generation as well as measures of ℓ_p balls. In Sect. 3, we define ℓ_p radially symmetric random vectors and provide basic results on ℓ_p radial and uniform distributions; see also the related concept of p-generakized normality [8]. In Sect. 4, we study the case $\mathbb{F} \equiv \mathbb{R}$ and present an algorithm for uniform generation of samples of the random vector $\mathbf{x}$ in ℓ_p balls. The algorithm consists of two main steps. In the first step, we generate samples of the components of $\mathbf{x}$ according to the so-called generalized gamma density [14]. Then, we compute $\mathbf{y} = r\mathbf{w}^{1/n}\mathbf{x}/\|\mathbf{x}\|_p$, where $\mathbf{w}$ is a real random variable uniformly distributed in $[0, 1]$. In Corollary 2, we prove that the samples $\mathbf{y}$ generated according to the above algorithm are indeed uniformly distributed in $B(r)$. In Sect. 5, we study the case $\mathbb{F} \equiv \mathbb{C}$ and we propose a similar algorithm for uniform generation in the complex ball $B(r)$. In Sect. 6 we discuss the generation of random matrix samples in Hilbert-Schmidt and induced norm balls. Finally, in Sect. 7 we show three examples of sample generation in real and complex ℓ_p balls. A preliminary version of this paper appeared in [4].

2   Definitions and Notation

2.1   Probability Densities and Random Sample Generation

In this subsection, we define two probability density functions that are used throughout the paper. We also recall a standard method for generating samples according to a given univariate distribution. In the sequel, the notation $\mathbf{x} \sim f$ means that $\mathbf{x}$ is a random element with probability density f.

Gamma density: A random variable $\mathbf{x} \in \mathbb{R}$ is gamma distributed with parameters (a, b), $\mathbf{x} \sim G(a, b)$, when it has density function

$$f_{\mathbf{x}}(x) = \frac{1}{\Gamma(a)b^a} x^{a-1} e^{-x/b}, \quad x \geq 0.$$

Efficient algorithms for the generation of gamma distributed random variables are described for instance in [13] and are available in most statistical packages, such as the Matlab Statistical Toolbox. A related density function is the so-called generalized gamma density [14].

Generalized Gamma density: A random variable $\mathbf{x} \in \mathbb{R}$ is generalized gamma distributed with parameters (a, c), $\mathbf{x} \sim \bar{G}(a, c)$, when it has density function

$$f_{\mathbf{x}}(x) = \frac{c}{\Gamma(a)} x^{ca-1} e^{-x^c}, \quad x \geq 0.$$

If a random generator for $G(a, b)$ is available, a random variable $\mathbf{x} \sim \bar{G}(a, c)$ can be simply obtained as $\mathbf{x} = \mathbf{z}^{1/c}$, where $\mathbf{z} \sim G(a, 1)$. This fact can be easily proved by means of a change of variable; see e.g. [7] for details.

A standard method for generating a univariate random variable with a given density function f, or with distribution function F, is the well-known *inversion method* [7,11] which is stated below.

Inversion method. Let $\mathbf{w} \in \mathbb{R}$ be a random variable with uniform distribution in $[0, 1]$. Let F be a continuous distribution function on $\mathbb{R}$ with inverse F^{-1} defined by

$$F^{-1}(u) = \inf\{x : F(x) = u, \ 0 < u < 1\}.$$

Then, the random variable $\mathbf{z} = F^{-1}(\mathbf{w})$ has distribution function F.

2.2   Measures of ℓ_p Balls

Let $\mathbb{F}$ be either the real field $\mathbb{R}$ or the complex field $\mathbb{C}$, and let $\|x\|_p = (\sum_{i=1}^{n} |x_i|^p)^{1/p}$, $p \in [1, \infty]$, be the standard ℓ_p norm. Then,

$$B(r) \doteq \{x \in \mathbb{F}^n : \|x\|_p \leq r\},$$

$$S(r) \doteq \{x \in \mathbb{F}^n : \|x\|_p = r\}$$

are the d-dimensional ℓ_p ball and its boundary, respectively. Clearly, the dimension d of $\mathbb{F}^n$ is n when $\mathbb{F} = \mathbb{R}$ and $2n$ when $\mathbb{F} = \mathbb{C}$. We denote as $\mu[B(r)]$ the d-dimensional measure (volume) of $B(r)$, defined as

$$\mu[B(r)] = \int_{B(r)} dV. \tag{1}$$

Similarly, $\mu[S(r)]$ is the $(d-1)$-dimensional measure (surface) of $S(r)$, defined as

$$\mu[S(r)] = \int_{S(r)} dV.$$

Clearly, $\mu[B(r)] = \mu[B(1)]r^d$ and $\mu[S(r)] = \mu[S(1)]r^{d-1}$. For radially defined sets as $B(r)$ the volume can be computed using a radially defined infinitesimal element of volume $dV = \nu\mu[S(r)]dr$, where ν is a constant independent of r but function of d and the ℓ_p norm used. Therefore, (1) becomes

$$\mu[B(r)] = \int_0^r \nu\mu[S(1)]\rho^{d-1}d\rho \qquad (2)$$

from which it follows that

$$\mu[B(r)] = \nu\mu[S(1)]\frac{r^d}{d}, \quad \mu[B(1)] = \nu\frac{\mu[S(1)]}{d}. \qquad (3)$$

3   Radial Density Functions

In this section we introduce radial symmetric density functions and state some basic results for this class of densities.

Definition 1. A random vector $\mathbf{x} \in \mathbb{F}^n$ is ℓ_p radially symmetric if its density function can be written as

$$f_\mathbf{x}(x) = g(r), \; r = \|x\|_p \qquad (4)$$

where $g(r)$ is called the *defining function* of $\mathbf{x}$.

In other words, for radially symmetric random vectors the density function is uniquely determined by its radial shape, which is described by $g(r)$. For given r, the level set of the density function is an equal probability set represented by $S(r)$.

Examples of radial densities for the real case. The classical multivariate normal density [11] with identity covariance matrix and mean value equal to zero is a radial density in ℓ_2 norm. In fact, setting $r = \|x\|_2$, we can write

$$f_\mathbf{x}(x) = \frac{1}{\sqrt{(2\pi)^n}}e^{-\frac{1}{2}x^T x} = \frac{1}{\sqrt{(2\pi)^n}}e^{-\frac{1}{2}r^2} = g(r).$$

The so-called multivariate Laplace density with zero mean value is a radial density in ℓ_1 norm. Indeed, setting $r = \|x\|_1$, we have

$$f_\mathbf{x}(x) = \frac{1}{2^n}e^{-\sum_{i=1}^n |x_i|} = \frac{1}{2^n}e^{-r} = g(r).$$

For a radially symmetric random vector $\mathbf{x}$ it is of interest to study how the random variable $\mathbf{r} = \|\mathbf{x}\|_p$ is distributed. This is analyzed in the next lemma.

Lemma 1. *If* $\mathbf{x}$ *is* ℓ_p *radially symmetric, then the random variable* $\mathbf{r} = \|\mathbf{x}\|_p$ *has density function* $f_\mathbf{r}(r)$ *given by*

$$f_\mathbf{r}(r) = \mu[B(1)]r^{d-1}g(r)d \qquad (5)$$

where $f_\mathbf{r}(r)$ *is called the norm density function.*

Proof. Notice that

$$F_r(r) = \text{Prob}\{\|x\|_p \leq r\} = \int_{B(r)} f_{\mathbf{x}}(x)dx = \int_{B(r)} g(\|x\|_p)dx.$$

Letting $r = \|x\|_p$ and performing a radial integration as in (2),

$$F_r(r) = \int_0^r \nu\mu[S(1)]\rho^{d-1}g(\rho)d\rho.$$

Differentiating with respect to r and using (3), we obtain (5). $\qquad\square$

The next lemma relates ℓ_p radially symmetric random vectors and uniform distributions on the level sets $S(r)$.

Lemma 2. *If* $\mathbf{x} \in \mathbb{F}^n$ *is* ℓ_p *radially symmetric with density function* $f_{\mathbf{x}}(x) = g(r)$, $r = \|x\|_p$, *then*
a) The conditional density $f_{\mathbf{x}|\mathbf{r}}$ *of* $\mathbf{x}$ *given* $\|\mathbf{x}\|_p = \mathbf{r}$ *is given by*

$$f_{\mathbf{x}|\mathbf{r}}(x|r) = \frac{1}{\mu[S(r)]}. \tag{6}$$

That is, $\mathbf{x}$ *is uniformly distributed on* $S(r)$.
b) The random vector $\mathbf{y} = \frac{\mathbf{x}}{\|\mathbf{x}\|_p}$ *is uniformly distributed on* $S(1)$.

Proof. a) If the random vector $\mathbf{x}$ is ℓ_p radially symmetric, then the conditional density is

$$f_{\mathbf{x}|\mathbf{r}}(x|r) = g(r)$$

and, consequently, $f_{\mathbf{x}|\mathbf{r}}$ is constant for given r. Since $f_{\mathbf{x}|\mathbf{r}}$ is a pdf

$$\int_{S(r)} f_{\mathbf{x}|\mathbf{r}}dV = f_{\mathbf{x}|\mathbf{r}} \int_{S(r)} dV = f_{\mathbf{x}|\mathbf{r}}\mu[S(r)] = 1$$

and this proves (6).
b) Clearly, $y = x/\|x\|_p \in S(1)$ for all x and we need to show that $\mathbf{y}$ is uniformly distributed on $S(1)$. Let $f_{\mathbf{y},\mathbf{r}}$ be the joint pdf of $\mathbf{y}$ and $\mathbf{r} = \|\mathbf{x}\|_p$, and $f_{\mathbf{y}|\mathbf{r}}$ the conditional pdf of $\mathbf{y}$ given $\mathbf{r}$, then

$$f_{\mathbf{y},\mathbf{r}}(y,r) = f_{\mathbf{y}|\mathbf{r}}(y|r)f_{\mathbf{r}}(r). \tag{7}$$

Writing $f_{\mathbf{y}|\mathbf{r}}$ as

$$f_{\mathbf{y}|\mathbf{r}}(y|r) = f_{\mathbf{y}|\mathbf{r}}\left(\frac{x}{\|x\|_p} \mid \|x\|_p = r\right) = f_{\mathbf{x}|\mathbf{r}}\left(\frac{x}{r} \mid \|x\|_p = r\right),$$

with the change of variable $z = x/r$ and using a)

$$f_{\mathbf{y}|\mathbf{r}}(y|r) = f_{\mathbf{z}|\mathbf{1}}(z \mid \|z\|_p = 1) = \frac{1}{\mu[S(1)]}.$$

Substituting in (7) and using (3) and (5), we obtain

$$f_{\mathbf{y},\mathbf{r}}(y,r) = \frac{1}{\mu[S(1)]}\mu[B(1)]r^{d-1}g(r)d = \nu r^{d-1}g(r).$$

Integrating with respect to r, we obtain the marginal density

$$f_{\mathbf{y}}(y) = \int_0^\infty \nu\rho^{d-1}g(\rho)d\rho.$$

Consider the norm density function $f_{\mathbf{r}}(r)$ in (5) and observe that its integral over $[0,\infty]$ is equal to one. Therefore

$$\int_0^\infty \rho^{d-1}g(\rho)d\rho = \frac{1}{\mu[B(1)]d}.$$

From this relation and (3) we finally obtain

$$f_{\mathbf{y}}(y) = \frac{\nu}{\mu[B(1)]d} = \frac{1}{\mu[S(1)]}.$$

□

We are now ready to state the key result of this section which shows how uniform distributions can be obtained from radially symmetric distributions and vice versa.

Theorem 1. *The following two conditions are equivalent:*

a) $\mathbf{x} \in \mathbb{F}^n$ *is ℓ_p radially symmetric with norm density function* $f_{\mathbf{r}}(r) = r^{d-1}d$, $r \in [0,1]$.
b) $\mathbf{x} \in \mathbb{F}^n$ *is uniformly distributed in $B(1)$.*

Proof. a)→b). Since $\mathbf{x}$ is ℓ_p radially symmetric, its norm density is given by (5), then

$$g(r) = \frac{1}{\mu[B(1)]} = f_{\mathbf{x}}(x).$$

Therefore, $\mathbf{x}$ is uniformly distributed in $B(1)$.
b)→a). Since $\mathbf{x}$ is uniform in $B(1)$ then

$$f_{\mathbf{x}}(x) = \begin{cases} \frac{1}{\mu[B(1)]} & \text{if } \|x\|_p \leq 1; \\ 0 & \text{otherwise.} \end{cases}$$

Notice that $f_{\mathbf{x}}(x)$ depends only on $\|x\|_p$; therefore $\mathbf{x}$ is ℓ_p radially symmetric with defining function $g(r) = f_{\mathbf{x}}(x)$, $r = \|x\|_p$. Substituting in (5) the claim is proved.

□

The next corollary shows how to obtain an ℓ_p radial distribution with given defining function, or with given norm density, starting from any arbitrary ℓ_p radial distribution.

Corollary 1. *Let* $\mathbf{x} \in \mathbb{F}^n$ *be* ℓ_p *radially symmetric and let* $\mathbf{z} \in \mathbb{R}^+$ *be an independent random variable with density* $f_{\mathbf{z}}(z)$, *then*

$$\mathbf{y} = \mathbf{z}\frac{\mathbf{x}}{\|\mathbf{x}\|_p}$$

is ℓ_p *radially symmetric and has norm density function* $f_{\mathbf{r}}(r) = f_{\mathbf{z}}(r)$, $r = \|y\|_p$. *The defining function* $g(r)$ *is consequently given by*

$$g(r) = \frac{1}{\mu[B(1)]r^{d-1}d}f_{\mathbf{r}}(r). \tag{8}$$

Proof. Clearly, $\mathbf{y}$ is ℓ_p radially symmetric and $\|\mathbf{y}\|_p = \mathbf{z}$. Therefore, the norm density function $f_{\mathbf{r}}(r)$ of $\mathbf{y}$ coincides with the density function $f_{\mathbf{z}}(z)$ of $\mathbf{z}$. Relation (8) follows immediately from (5). □

The previous corollary may be used to generate ℓ_p radially symmetric random vectors with given defining functions. In the next corollary, we specialize this result to uniform distributions.

Corollary 2. *Let* $\mathbf{x} \in \mathbb{F}^n$ *be* ℓ_p *radially symmetric and let* $\mathbf{w} \in \mathbb{R}$ *be an independent random variable uniformly distributed in* $[0,1]$. *Then,*

$$\mathbf{y} = r\mathbf{z}\frac{\mathbf{x}}{\|\mathbf{x}\|_p}, \quad \mathbf{z} = \mathbf{w}^{1/d}$$

is uniformly distributed in $B(r)$.

Proof. By the inversion method, it follows that the distribution of $\mathbf{z}$ is $F_{\mathbf{z}}(z) = z^d$, therefore $f_{\mathbf{z}}(z) = dz^{d-1}$. For $r = 1$, the statement is immediately proved by means of Theorem 1. With a rescaling, it is immediate to show that $\mathbf{y}$ is uniformly distributed in $B(r)$. □

This result can be interpreted as follows: First, an ℓ_p radially symmetric random vector $\mathbf{x}$ is normalized to obtain a uniform distribution on the surface $S(r)$ of the set $B(r)$, then each sample is forced into $B(r)$ by the *volumetric factor* $\mathbf{z}$. Therefore, the problem of uniform generation is reduced to that of generation of ℓ_p radially symmetric random vectors. This is discussed in the next section.

4   Generation of Uniform Random Real Vectors

In this section, we propose an algorithm to generate real vectors $\mathbf{x} \in \mathbb{R}^n$ uniformly distributed in ℓ_p balls. This algorithm is based on the results of the previous sections and, in particular, on Corollary 2. Let $\mathbf{x} = [\mathbf{x}_1 \ \mathbf{x}_2 \ \cdots \mathbf{x}_n]^T$, where $\mathbf{x}_i \in \mathbb{R}$ are independent identically distributed (i.i.d) random variables with density function

$$f_{\mathbf{x}_i}(x_i) = k_R e^{-|x_i|^p}, \quad k_R = \frac{p}{2\Gamma(\frac{1}{p})}. \tag{9}$$

Notice that the density function (9) is a bilateral generalized gamma density with parameters $(\frac{1}{p}, p)$. In particular, for $p = 1$ equation (9) is the Laplace density and for $p = 2$ it is the well-known normal density with zero mean value and variance equal to $1/2$.

Since the $\mathbf{x}_i$'s are independent, the joint density $f_{\mathbf{x}}(x)$ can be written as

$$f_{\mathbf{x}}(x) = \prod_{i=1}^{n} k_R e^{-|x_i|^p} = k_R^n e^{-\|x\|_p^p} = g(r), \; r = \|x\|_p.$$

From the above expression, we notice that $\mathbf{x}$ is ℓ_p radially symmetric and therefore Corollary 2 can be immediately applied. In addition, since $f_{\mathbf{x}}(x)$ is a density function, its integral is equal to one. If we evaluate this integral using a radially infinitesimal element, we obtain

$$\int_0^\infty k_R^n \nu \mu[S(1)] \rho^{n-1} e^{-\rho^p} d\rho = k_R^n \nu \mu[S(1)] \frac{\Gamma(\frac{n}{p})}{p} = 1.$$

Next, using (3), a closed form relation for the volume of the ℓ_p ball can be obtained

$$\mu[B(r)] = \mu[B(1)]r^n, \;\; \mu[B(1)] = 2^n \frac{\Gamma^n(\frac{1}{p} + 1)}{\Gamma(\frac{n}{p} + 1)}.$$

This formula may be used for computing the acceptance rate of a rejection method [7,13]. To explain this method, one first observes that the set $B(r)$ can be overbounded with the n-dimensional hypercube $\{x \in \mathbb{R}^n : \|x\|_\infty \leq r\}$. A uniform distribution in the hypercube is easily obtained generating independently each component of the vector $\mathbf{x}$ uniformly in $[-r, r]$. Finally, one rejects the samples which are outside the ℓ_p ball. Clearly, the acceptance rate R_a is equal to the ratio of the volume of the two balls

$$R_a = \frac{\Gamma^n(\frac{1}{p} + 1)}{\Gamma(\frac{n}{p} + 1)}.$$

This formula immediately shows the inefficacy of the rejection method for large n, see Table 1.

We are now ready to present the algorithm for uniform sample generation in real ℓ_p balls.

4.1 Algorithm for Real Uniform Generation

Given n, p, and r, the algorithm returns a real random vector $\mathbf{y}$ which is uniformly distributed in $B(r)$.

1. Generate n independent random real scalars $\xi_i \sim \bar{G}(\frac{1}{p}, p)$.
2. Construct the vector $\mathbf{x} \in \mathbb{R}^n$ of components $\mathbf{x}_i = s_i \xi_i$, where s_i are independent random signs.
3. Generate $z = w^{1/n}$, where w is a random variable uniformly distributed in the interval $[0, 1]$.
4. Return $\mathbf{y} = rz \frac{\mathbf{x}}{\|\mathbf{x}\|_p}$.

Table 1. Acceptance rate of the rejection algorithm for the real case.

	$p = 1$	$p = 1.5$	$p = 2$
$n = 2$	0.5	0.68	0.79
$n = 3$	0.17	0.37	0.52
$n = 4$	0.042	0.17	0.31
$n = 5$	8.33e-3	6.47e-2	0.16
$n = 10$	2.76e-7	1.39e-4	2.49e-3
$n = 20$	4.11e-19	8.68e-12	2.46e-8
$n = 30$	3.77e-33	1.91e-20	2.04e-14

5   Generation of Uniform Random Complex Vectors

In this section, we propose an algorithm based on Corollary 2 to generate complex vectors $\mathbf{x} \in \mathbb{C}^n$ uniformly distributed in ℓ_p balls. Let $\mathbf{x} = [x_1 \; x_2 \; \cdots x_n]^T$, where $x_i \in \mathbb{C}$. Each component x_i can be considered as a two-dimensional real vector $\bar{\mathbf{x}}_i \in \mathbb{R}^2$ formed by its real and imaginary parts, so that the absolute value of x_i coincides with the ℓ_2 norm of $\bar{\mathbf{x}}_i$. Let each $\bar{\mathbf{x}}_i$ be an i.i.d. ℓ_2 radially symmetric vector with defining function

$$f_{\bar{\mathbf{x}}_i}(\bar{x}_i) = k_C e^{-r_i^p} = g_i(r_i), \quad r_i = \|\bar{\mathbf{x}}_i\|_2,$$

where $k_C = \frac{p}{2\pi\Gamma(\frac{2}{p})}$. Therefore, using (5), the corresponding norm density function is

$$f_{\mathbf{r}_i}(r_i) = 2\pi k_C r_i e^{-r_i^p}, \quad r_i \geq 0. \tag{10}$$

The above density is indeed a generalized gamma density with parameters $(\frac{2}{p}, p)$. The density function $f_{\mathbf{x}}(x)$ is the joint density function of the x_i's. Since the x_i's are independent

$$f_{\mathbf{x}}(x) = \prod_{i=1}^{n} k_C e^{-|x_i|^p} = k_C^n e^{-\|x\|_p^p} = g(r), \quad r = \|x\|_p.$$

As in the real case, from the above expression it follows that the random vector $\mathbf{x}$ is ℓ_p radially symmetric and the results of Corollary 2 can be applied. Following the same considerations of Sect. 4, we obtain a closed form relation for the volume of the complex ℓ_p ball

$$\mu[B(r)] = \mu[B(1)]r^{2n}, \quad \mu[B(1)] = \pi^n \frac{\Gamma^n(\frac{2}{p} + 1)}{\Gamma(\frac{2n}{p} + 1)}.$$

In this case, a rejection method can be based on the complex hypercube $\{x \in \mathbb{C}^n : \|x\|_\infty \leq r\}$. A random vector $\mathbf{x}$ uniformly distributed in this set is trivially obtained generating independently each component x_i uniformly in the complex disk of radius r. The acceptance rate R_a of this method is given by

$$R_a = \frac{\Gamma^n(\frac{2}{p} + 1)}{\Gamma(\frac{2n}{p} + 1)}.$$

In Table 2 several values of R_a are reported, showing the inefficiency of the rejection method for large n.

Table 2. Acceptance rate of the rejection algorithm for the complex case.

	$p = 1$	$p = 1.5$	$p = 2$
$n = 2$	0.17	0.35	0.5
$n = 3$	0.011	0.070	0.17
$n = 4$	3.97e-4	9.39e-3	4.17e-2
$n = 5$	8.82e-6	9.23e-4	8.33e-3
$n = 10$	4.21e-16	3.85e-10	2.76e-7
$n = 20$	1.29e-42	9.07e-27	4.11e-19
$n = 30$	1.29e-73	2.30e-46	3.77e-33

5.1 Algorithm for Complex Uniform Generation

Given n, p, and r, the algorithm returns a complex random vector $\mathbf{y}$ with uniform distribution in $B(r)$.

1. Generate n independent complex numbers $\xi_i = e^{j\theta}$, where θ is uniform in $[0, 2\pi]$ (the ξ_i's are uniformly distributed on the complex unit circle.)
2. Construct the vector $\mathbf{x} \in \mathbb{C}^n$ of components $x_i = \eta_i \xi_i$, where the η_i's are independent random variables $\eta_i \sim \bar{G}(\frac{2}{p}, p)$.
3. Generate $z = w^{1/(2n)}$, where w is uniformly distributed in $[0, 1]$.
4. Return $\mathbf{y} = rz \frac{\mathbf{x}}{\|\mathbf{x}\|_p}$.

6 Random Matrices

In this section we discuss the uniform generation of matrix samples in various norm balls. We use the same notation as for the vector case, denoting the norm ball of radius r as

$$B(r) = \{X \in \mathbb{F}^{n,m} : \|X\| \le r\}.$$

A random matrix $X \in \mathbb{F}^{n,m}$ is uniformly distributed in $B(r)$ if the joint probability density function of its elements is given by

$$f_X(X) \doteq \begin{cases} \text{constant} & \text{if } X \in B(r); \\ 0 & \text{otherwise.} \end{cases}$$

The cases of Hilbert-Schmidt and operator induced norms will be considered in the sequel.

6.1 Hilbert-Schmidt Norms

Hilbert-Schmidt norms are based on the isomorphism between $\mathbb{F}^{n,m}$ and $\mathbb{F}^{nm}$. Let $\Gamma : \mathbb{F}^{n,m} \to \mathbb{F}^{nm}$ be a column vectorization operator, then the Hilbert-Schmidt p-norm on $X \in \mathbb{F}^{n,m}$ is defined as

$$\|X\|_{H,p} \doteq \|\Gamma(X)\|_p.$$

Therefore, the generation of matrix samples in these norms is equivalent to the vector case considered in the previous sections. We remark that for $p = 2$ the Hilbert-Schmidt norm is in fact the frequently used Frobenius matrix norm.

6.2 Induced Norms

For any vector p-norm, the induced operator norm (gain) on $X \in \mathbb{F}^{n,m}$ is defined as

$$\|X\|_{p-ind} \doteq \max_{\|\xi\|_p=1} \|X\xi\|_p, \quad \xi \in \mathbb{F}^m.$$

We will discuss here the cases with $p = 1, \infty$ and $p = 2$. For $p = 1$ we have that (see e.g. [9])

$$\|X\|_{1-ind} = \max_{1\le j\le m} \sum_{i=1}^{n} |x_{ij}|.$$

Writing X by columns as $X = [x_1 \ x_2 \ \cdots \ x_m]$, the 1-induced norm is the maximum of the 1-norms of the columns of X

$$\|X\|_{1-ind} = \max_{1\le j\le m} \|x_j\|_1.$$

Assume that each column vector $x_j \in \mathbb{F}^n$ is independently generated uniformly in the vector 1-norm ball of radius r. That is, for $1 \le j \le m$

$$f_x(x_j) = \begin{cases} K \ \text{if } \|x_j\|_1 \le r; \\ 0 \ \text{otherwise.} \end{cases}$$

As the columns of X are generated independently, the probability density of X may be written as

$$f_X(X) = f_X(x_1, x_2, \dots, x_m) = f_x(x_1)f_x(x_2)\cdots f_x(x_m).$$

Therefore $f_X(X)$ is zero whenever $f_x(x_j) = 0$ for some x_j and it is constant when $f_x(x_j) = K$ for all x_j. This means that

$$f_X(X) = \begin{cases} \text{constant} \ \text{if } \max_{1\le j\le m} \|x_j\|_1 \le r; \\ 0 \qquad\quad \text{otherwise} \end{cases}$$

which is precisely the definition of a uniform density in the matrix 1-induced norm ball of radius r.

Summarizing, uniform matrix samples in the 1-induced norm ball of radius r are obtained by generating independently each column x_j of X with uniform distribution in the (vector) 1-norm ball of radius r.

The case with $p = \infty$ is equivalent to the previous case. Expressing X by rows $X = [x_1 \ x_2; \cdots \ x_n]^T$, we have that [9]

$$\|X\|_{\infty-ind} = \max_{1 \le i \le n} \|x_i^T\|_1.$$

That is, the ∞-induced norm is the maximum of the 1-norms of the rows of X.

Using the same arguments as before, we have that uniform matrix samples in the ∞-induced norm ball of radius r are obtained by generating independently each row x_i^T of X with uniform distribution in the (vector) 1-norm ball of radius r.

The case with $p = 2$ has a particular relevance in control theory; see e.g. [19]. It is well-known that the 2-induced norm of a matrix is equal to its maximum singular value

$$\|X\|_{2-ind} = \bar{\sigma}(X)$$

and it is often referred to as the *spectral* matrix norm. Unfortunately, the problem of uniform generation in the spectral norm is much more difficult than the previously considered cases. In fact, in the previous cases, the generation of matrix samples is reduced to the generation of vector samples. This is not the case for the spectral norm. The sample generation in the spectral norm is studied in [5], [6].

6.3   Radially Distributed Matrices

We remark that the results of Sect. 3 hold for the matrix case with minor technical modifications. In particular, the result of Corollary 1 can be applied to generate matrices with a generic radial distribution on the support $B(r)$. As an example, to generate complex $n \times n$ random matrices Y with a truncated gaussian (unilateral) norm density on the support $B(1) = \{\|X\|_{H,2} \le 1\}$ (Frobenius norm-ball of unit radius), we generate uniform matrix samples X as discussed in Subsect. 6.1, and then rescale the samples as

$$Y = z \frac{X}{\|X\|_{H,2}}$$

where z is an independent scalar random variable with the desired truncated gaussian density.

7   Examples

In this section, we show three examples of uniform generation of vector samples in various sets. For illustrative purposes, we consider the case $n = 2$. In the first example, we take $p = 1.5$ and generate $N = 5,000$ samples of real two-dimensional vectors uniformly distributed in $B(1)$, using the algorithm presented in Subsect. 4.1. Fig. 1 shows the sample generation.

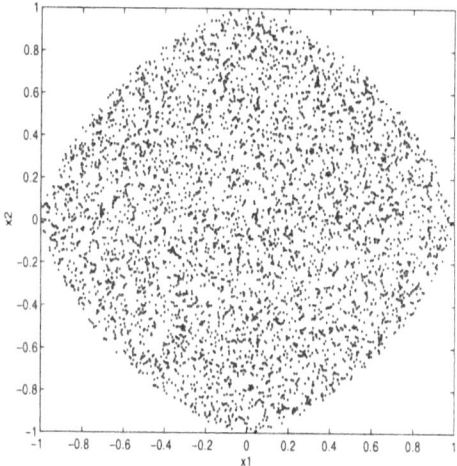

Fig. 1. Two dimensional real vectors uniformly distributed in $B(1)$, for $p = 1.5$.

We remark that the algorithm of Subsection 4.1 can be also used when $p \in (0, 1)$. However, in this case $\|x\|_p^p = \sum_{i=1}^n |x_i|^p$ is not a norm and the set

$$B(r) = \{x \in \mathbb{R}^n : \|x\|_p \leq r\}$$

is not convex. In Fig. 2, we show the generation of $N = 5,000$ samples of real two-dimensional vectors uniformly distributed in $R(1)$ for $p = 0.7$.
Finally, we consider the generation of $N = 5,000$ samples of vectors $x \in \mathbb{C}^2$ uniformly distributed in $B(1)$ for $p = 1$. Notice that in this case the samples belong to a linear space of dimension $d = 4$, therefore in Fig. 3 we show two-dimensional projections of the samples.

8   Conclusions

In this paper, we studied uniform sample generation in the ball $B(r) = \{x \in \mathbb{F}^n : \|x\|_p \leq r\}$, where $\mathbb{F}$ is either the real or the complex field. We have shown that this task can be performed by means of simple algorithms which use the generalized gamma density. In Corollary 2, we proved that the samples generated in this way are indeed uniformly distributed. Also, the basics for generation of ℓ_p radially symmetric random vectors with desired defining function, not necessarly uniform, have been provided in Corollary 1. The extensions to matrix samples generation in Hilbert-Schmidt and $1, \infty$-induced norms is discussed in Sect. 6.

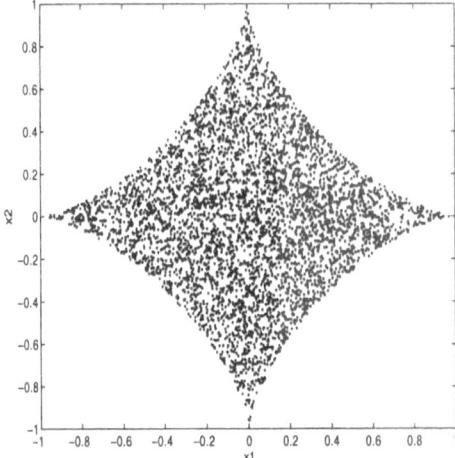

Fig. 2. Two dimensional real vectors uniformly distributed in $B(1)$, for $p = 0.7$.

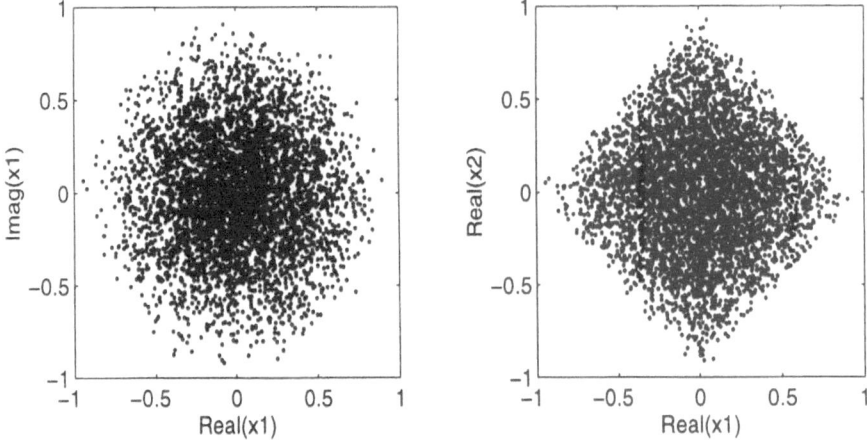

Fig. 3. Two-dimensional projections of complex vectors $x = [x_1, x_2]^T$, $x_1, x_2 \in \mathbb{C}$, uniformly distributed in $B(1)$, for $p = 1$.

The proposed algorithms have been implemented in Matlab and are available on the web at the address www.polito.it/~dabbene/uniform.html.

References

1. E.W. Bai, R. Tempo and M. Fu, "Worst case properties of the uniform distribution and randomized algorithms for robustness analysis," to appear in *Mathematics of Control, Signals, and Systems*, 1998.
2. B.R. Barmish and C.M. Lagoa, "The uniform distribution: a rigorous justification for its use in robustness analysis," *Mathematics of Control, Signals, and Systems*, vol. 10, pp. 203–222, 1997.

3. B.R. Barmish, C.M. Lagoa and R. Tempo, "Radially truncated uniform distribution for probabilistic robustness of control systems," in *Proc. of Amer. Cont. Conf.*, Albuquerque, New Mexico, 1997.
4. G. Calafiore, F. Dabbene and R. Tempo, "Uniform sample generation in ℓ_p balls for probabilistic robustness analysis," in *Proc. of the Conf. on Dec. and Cont.*, Tampa, Florida, 1998.
5. G. Calafiore, F. Dabbene and R. Tempo, "The probabilistic real stability radius," Politecnico di Torino, CENS-CNR Technical Report 98-2a, 1998.
6. G. Calafiore, F. Dabbene and R. Tempo, "Uniform distributions in the spectral norm," Politecnico di Torino, CENS-CNR Technical Report 98-2b, 1998.
7. L. Devroye, *Non-Uniform Random Variate Generation*. Springer-Verlag, New York, 1986.
8. A.K. Gupta, "Characterization of p-generalized normality," *Journal of Multivariate Analysis*, vol. 60, pp. 61–71, 1997.
9. R.A. Horn and C.R. Johnson, *Matrix Analysis*, Cambridge University Press, 1985.
10. P.P. Khargonekar and A. Tikku, "Randomized algorithms for robust control analysis have polynomial time complexity," in *Proc. of the Conf. on Dec. and Cont.*, Kobe, Japan, 1996.
11. A. Papoulis, *Probability, Random Variables, and Stochastic Processes*. McGraw-Hill, New York, 1965.
12. L.R. Ray and R.F. Stengel, "A Monte Carlo approach to the analysis of control system Robustness," *Automatica*, vol. 29, pp. 229–236, 1993.
13. R.Y. Rubinstein, *Simulation and the Monte Carlo Method*. Wiley, New York, 1981.
14. E.W. Stacy, "A Generalization of the Gamma distribution," *Annals of Mathematical Statistics*, vol. 33, pp. 1187–1192, 1962.
15. R.F. Stengel and L.R. Ray, "Stochastic robustness of linear time-invariant control systems," *IEEE Trans. on Aut. Cont.*, vol. 36, pp. 82–87, 1991.
16. R. Tempo, E.W. Bai and F. Dabbene, "Probabilistic robustness analysis: explicit bounds for the minimum number of samples," *Sys. and Cont. Lett.*, vol. 30, pp. 237–242, 1997.
17. M. Vidyasagar, *A Theory of Learning and Generalization with Applications to Neural Networks and Control Systems*, Springer-Verlag, Berlin, 1996.
18. M. Vidyasagar, "Statistical learning theory: an introduction and applications to randomized algorithms," in *Proc. of the Europ. Cont. Conf.*, Brussels, Belgium, 1997.
19. K. Zhou, J.C. Doyle, K. Glover, *Robust and Optimal Control*, Prentice-Hall, Upper Saddler River, 1996.

Decentralized Reliable Control for Large-Scale Systems

Zhengfang Chen and Timothy Chang

Department of Electrical and Computer Engineering, New Jersey Institute of
Technology, University Heighs, Newark, New Jersey 07102 USA
Email: zhengfangchen@lucent.com, changtn@admin.njit.edu

Abstract. In this paper, the reliable control for large-scale systems is considered. Reliable control concerns the ability of closed loop system to maintain stability and regulation properties during arbitrary sensor, controller, and actuator failure without being retuned. Two concepts of reliable control are introduced in this work: (1) the Decentralized Robust Servomechanism Problem with Complete Reliability (DRSPwCR) and (2) the Block Decentralized Robust Servo Problem with Complete Reliability (BDRSPwCR). DRSPwCR deals with the reliable control problem of systems with a strict diagonal decentralized controller configuration. It is shown that DRSPwCR is solvable for "reliable" systems whose steady state gain matrices belong to the class of P matrices and for "unreliable", minimum phase systems by applying strict decentralized polynomial compensation to the plant and thereby extending the class of processes that can be controlled reliably. For plants which have a pre-imposed block diagonal structure or non-minimum phase minors, BDRSPwCR is defined and solved. In this case, a matrix polynomial compensation is applied to the plant to achieve complete reliability against arbitrary sensor/actuator/controller failure.

1 Background and Introduction

Reliable control concerns the ability of closed loop system to maintain key properties such as stability and regulation during arbitrary sensor, actuator, or partial controller failure. In many industrial control problems, reliability requirements are critical to the long term feasibility of system operations.

Existing approaches in dealing with partial system failure can be classified as 1) fault-sensing and 2) fault-tolerant. The latter is generally considered as "reliable" control where the system with partial failure can remain maximally functional without retuning the controllers. In the past 10 years, a number of methods have emerged in the research of reliable control: extended $\mathcal{H}_\infty$ [9], algebraic factorizations [8] design, and robust servo control [1] design, etc. The $\mathcal{H}_\infty$-norm has long been used as a performance measure in solving diverse control problems including disturbance rejection, model reference design, tracking and robust design. The extended $\mathcal{H}_\infty$ design method for the reliable centralized and decentralized control systems was observer-based, resulting in designs with guaranteed closed-loop stability and an $\mathcal{H}_\infty$ disturbance-attenuation bound for the nominal plant and the occurrence of pre-determined sensor/actuator failure. The design is obtained by including in the nominal plant description of additional disturbance inputs or regulated inputs to account for possible control input or sensor outages, respectively, and computing basic $\mathcal{H}_\infty$ control designs for the augmented plant. The existence of solutions of the design equations are sufficient to guarantee that the reliable design tolerates system component outages within a prespecified set of susceptible sensors or actuators in the centralized case or within a prespecified set of susceptible control channels in the decentralized case. The algebraic factorization methodology is an algebraic design method which was developed based on the diagonalization of certain transfer functions of the nominal system which is linear, time-invariant, multi-input, multi-output with unity-feedback [5–8]. In this approach, the reliable control problem is treated as the simultaneous stabilization of the nominal plant and the plant multiplied by different failure matrices. The system is required to be stable for all possible failures of at most k of the sensor-connections or at most m of the actuator-connections. In this case the plant and the controller must have certain properties. These properties are explained in terms of denominator-matrices of their coprime-factorizations. A controller design methodology, which does not require the failure to be known in advance, is obtained by the diagonalization of certain transfer functions of the nominal system having the above properties. The robust servo control design methodology [1,2] is directed at large scale systems where it is generally difficult to obtain a precise mathematical model. The control synthesis is carried out by using decentralized design, i.e. the multi-variable system is treated as though it consists of a number of subsystems and the design effort is then focused on analyzing interaction among the subsystems. The only assumptions made for the plant are that (i) it can be described by a linear time-invariant model and (ii) it is open loop asymptotically stable. Under these assumptions, a set of steady-state interaction indices are applied to provide a measure of closed loop interaction and subsequently to determine if a certain system can be made fault-tolerant. This paper follows the robust servo control approach where the decentralized robust servomechanism problem with complete reliability is considered. The failure pattern considered in this work includes arbitrary sensor, actuator, and controller failure, thereby imposing the restriction that the system be open-loop stable. The primary objectives of the reliable controls are: 1) disturbance rejection and 2) stability of the failed system and regulation of the remaining system.

2 Development

The plant is represented by the following model:

$$\dot{x} = Ax + Bu + E\omega$$
$$y = Cx \tag{1}$$

where $A \in \mathbb{R}^{n \times n}$ is assumed to be stable, $B = [b_1 \ldots b_\nu] \in \mathbb{R}^{n \times \nu}$, $E \in \mathbb{R}^{n \times \Omega}$, and $C = [c'_1 \ldots c'_\nu]' \in \mathbb{R}^{\nu \times n}$. The variables $x \in \mathbb{R}^n$, $u = [u_1 \ldots u_\nu]' \in \mathbb{R}^\nu$, $y = [y_1 \ldots y_\nu]' \in \mathbb{R}^\nu$ and $\omega \in \mathbb{R}^\Omega$ are the state, input, output and the constant disturbance vectors, respectively.

The open loop transfer matrix can be expressed as:

$$T(s) = C(sI - A)^{-1}B = \frac{N(s)}{d(s)} = \{t_{ij}(s)\}; i, j = 1, 2, \ldots, \nu. \qquad (2)$$

Define the following normalized matrix:

$$\overline{T}(s) = T(s)[\text{diag}(t_{11}(s), t_{22}(s), \ldots, t_{\nu\nu}(s)]^{-1}. \qquad (3)$$

The matrix $\overline{T}[i_1, i_2, \ldots, i_\phi](s)$ is obtained from $\overline{T}(s)$ by retaining only the $i_1, i_2, \ldots, i_\phi$ non-redundant rows and columns where $i_1, i_2, \ldots, i_\phi \in [1, \nu]$, $\phi \in [1, \nu]$. Similarly, let $N[i_1, i_2, \ldots, i_\phi](s)$ be obtained from $N(s)$ by retaining $i_1, i_2, \ldots, i_\phi$ non-redundant rows and columns.

The set of ϕ-th orders leading principal minors of $N(s)$ is defined as:

$$\det N[\phi](s) = \{p(s) \in \mathbb{C} : p(s) = \det(N[i_1, i_2, \ldots, i_\phi](s))\} \qquad (4)$$

whose roots are just the transmission zeros of the $\phi \times \phi$ principal subsystems of (1).

With no loss of generality, it is assumed that the control structure is given by the following input/output pairing:

$$\{(u_1, y_1), (u_2, y_2), \ldots, (u_\nu, y_\nu)\}. \qquad (5)$$

Definition 1. [1] Decentralized Robust Servomechanism Problem (DRSP): Given the plant (1) and input/output pairing (5), obtain a decentralized controller using the pairing (5) so that the following conditions all hold:

1. The closed loop system is asymptotically stable.
2. Asymptotic tracking occurs, i.e. $\lim_{t \to \infty} y(t) = 0$, for all constant ω.
3. Property 2 holds for parametric perturbations: $A \to A + \delta A, B \to B + \delta B$, and $C \to C + \delta C$ provided that the closed loop system remains stable.

Lemma 1. [1] *There exists a solution to the DRSP with controller installation sequence* $\{i_1, i_2, \ldots, i_\nu\}$ [1] *iff the following conditions all hold:*

- $\det N[i_1](0) \neq 0, \forall i_1 \in \{1, 2, \ldots, \nu\}$
- $\det N[i_1, i_2](0) \neq 0, \forall i_1, i_2 \in \{1, 2, \ldots, \nu\}$
- $\det N[i_1, i_2, i_3](0) \neq 0, \forall i_1, i_2, i_3 \in \{1, 2, \ldots, \nu\}$

- $\vdots$
- $\det N(0) \neq 0.$

The following system failure cases are now defined:

Definition 2. Sensor Failure: The i-th sensor is said to have failed at time $t_1 > 0$ if

$$y_i(t) = 0 \quad \forall t > t_1.$$

The i-th sensor failure reflects the situation where a sensor in the i-th channel ceases to function and generates only a null output for all time thereafter.

Definition 3. Controller/Actuator Failure: The i-th controller/actuator is said to have failed at time $t_2 > 0$ if

$$u_i(t) = 0 \quad \forall t > t_2.$$

The i-th controller or actuator failure reflects the situation where a controller or actuator in i-th channel ceases to function and generates only a null output for all time thereafter.

Definition 4. [4] Plant with Multiple Failures: Given the plant (1) where $\nu - \phi$ sensors/actuators/controllers have failed so that the resulting plant is described by:

$$\dot{x} = Ax + b_{i_1} u_{i_1} + b_{i_2} u_{i_2} + \cdots + b_{i_\phi} u_{i_\phi}$$
$$y_i = c_i x, \quad i \in \{i_1, i_2, \ldots, i_\phi\}. \tag{6}$$

The ϕ-input/output plant given by (6) is said to be a plant with partial sensor, actuator, or controller failure.

Definition 4 reflects the situation where $\nu - \phi$ channels have effectively gone "off-line". The objective of reliable control is to maintain stability and regulation for (6) without retuning of the remaining "ϕ" controllers.

Definition 5. [4] DRSP with Complete Reliability (DRSPwCR): Given the plant (1) and the input/output pairing (5), obtain a decentralized controller so that the following conditions all hold:

1. There exists a solution to the DRSP for the normal plant (1).
2. Under arbitrary failure of $\nu - \phi$ sensor/actuator/control, where $\phi \in [1, \nu]$, the remaining ϕ controllers solve the DRSP for (6) without retuning.

The controller structure considered here can be described by the following equations:

$$\dot{\xi} = \Lambda_0^i \xi_i + \Lambda_1^i u_i + \Lambda_2^i y_i \tag{7}$$
$$\dot{\eta}_i = y_i \tag{8}$$
$$u_i = K_1^i y_i + K_2^i \xi_i + K^i \eta_i \tag{9}$$

where $i = 1, 2, \ldots, \nu$ and $\xi_i \in \mathbb{R}^{\kappa_i}$. The above decentralized controller (7–9), when expressed in frequency domain, can be described by the following equation:

$$C_i(s) = \frac{U_i(s)}{Y_i(s)} = \frac{k_{\kappa_i+1}^i s^{\kappa_i+1} + k_{\kappa_i}^i s^{\kappa_i} + \cdots + k_1^i s + k_0^i}{(d_\kappa s^\kappa + d_{\kappa-1} s^{\kappa-1} + \cdots + d_1 s + d_0)s} = \frac{n_{ci}(s)}{d_{ci}(s)} \tag{10}$$

where $i \in [1, \nu]$.

Lemma 2. [1] *There exists a solution to the DRSPwCR using the decentralized controller (7)–(9) if the following conditions all hold:*

- $\det \overline{T}[i_1, i_2](0) > 0, \ \forall i_1, i_2 \in \{1, 2, \ldots, \nu\}$
- $\det \overline{T}[i_1, i_2, i_3](0) > 0, \ \forall i_1, i_2, i_3 \in \{1, 2, \ldots, \nu\}$

- $\vdots$
- $\det \overline{T}(0) > 0.$

Remark 1. Lemma 2 requires that all principal minors of $\overline{T}(0)$ be strictly positive, a condition not always satisfied by an arbitrary plant.

The following theorem is obtained:

Theorem 1. *Given the plant (1), DRSPwCR can be achieved only if the controller is non-proper. Controller (7–9) is always proper and therefore it can not solve DRSPwCR.*

Proof. Consider the case of a 2-I/O system, let the transfer matrix of the system be

$$G(s) = \frac{1}{d(s)} \begin{bmatrix} n_{11}(s) & n_{12}(s) \\ n_{21}(s) & n_{22}(s) \end{bmatrix}$$

The closed loop characteristic polynomials are:

$$D_1(s) = d_{c1}(s)d(s) - n_{11}(s)n_{c1}(s) \tag{11}$$
$$D_2(s) = d_{c2}(s)d(s) - n_{22}(s)n_{c2}(s) \tag{12}$$
$$D_{12}(s) = D_1(s)D_2(s) - n_{12}(s)n_{21}(s)n_{c1}(s)n_{c2}(s) \tag{13}$$

where $D_1(s)$ is the closed loop system characteristic polynomial when controller 1 only is applied, similarly $D_2(s)$ is the closed loop system characteristic polynomial when controller 2 only is applied and $D_{12}(s)$ is the closed loop characteristic polynomial when both controllers are applied.

Now since the plant is assumed to be strictly proper,

- $\deg D_1(s) = \kappa + n + 1 > \deg n_{11}(s)n_{c1}(s) = \kappa + n.$
- $\deg D_2(s) = \kappa + n + 1 > \deg n_{22}(s)n_{c2}(s) = \kappa + n.$

Similarly, $\deg(D_1 D_2)(s) > \deg(n_{12}n_{21}n_{c1}n_{c2})(s)$, From (13), the degree of $D_{12}(s)$ will be the same as that of $(D_1 D_2)(s)$, and the highest order coefficients of $D_{12}(s)$ and $(D_1 D_2)(s)$ are the same.

Since $D_{12}(s)$ and $(D_1 D_2)(s)$ are both Hurwitz, therefore, their coefficients must necessarily be of same signs. For the lowest order terms,

$$D_{12}(0)D_1(0)D_2(0) > 0. \tag{14}$$

From the necessary requirement of reliability consideration:

$$D_1(0)D_2(0)D_{12}(0) < 0. \tag{15}$$

Proof of (15) is given in Appendix A. Equation (14) violates (15). This implies that the assumption is wrong and $m_i > n \ \forall i \in [1, \nu]$. $\qquad \square$

Remark 2. A plant satisfying all the conditions in Lemma 2 is called a "reliable system". Otherwise, it is referred to as an "unreliable system". To solve the DRSPwCR for an unreliable system, it is not adequate to use a controller of the form (7)–(9) only. Existing reliable control results based on the Robust Servo Control method are limited to reliable systems.

3   Reliable Control for Minimum Phase Systems

In this section, results for solving Decentralized Robust Servo Problem with Complete Reliability (DRSPwCR) for a class of open loop stable, minimum phase system are developed. The decentralized PID^r controller configuration is applied and simulation results of numerical examples are given.

3.1   Main Results

From the discussions in the previous section, it follows that when a given plant does not satisfy all conditions in Lemma 2, the control strategy (7)–(9) alone cannot solve the DRSPwCR. However, for a certain group of unreliable, minimum phase systems, the following new results are now obtained to relax the conditions of Lemma 2:

Theorem 2. *There exists a solution to the DRSPwCR (Definition 5) if the following conditions all hold:*

1. *There is a solution to DRSP for (1).*
2. $\det N[\phi](s), \phi = 1, 2, \ldots, \nu$, *do not possess non-minimum phase zeros.*

To solve the DRSPwCR, a possible controller is the PID^r type given below:

$$K_P(s) = \operatorname{diag}(k_{p1}, k_{p2}, \ldots, k_{p\nu}) \tag{16}$$

$$K_I(s) = \operatorname{diag}\left(\frac{k_1}{s}, \frac{k_2}{s}, \ldots, \frac{k_\nu}{s}\right) \tag{17}$$

$$k_i^d(s) = k_i^1 s + k_i^2 s^2 + \cdots + k_i^r s^r, i = 1, 2, \ldots, \nu \tag{18}$$

$$K_D(s) = \operatorname{diag}(k_1^d(s), k_2^d(s), \ldots, k_\nu^d(s)) \tag{19}$$

$$K_{PD}(s) = K_P(s) + K_D(s) \tag{20}$$

where r is the maximum pole-zero excess defined as:

$$r \triangleq n - \min\{\deg \det N(\phi) : \phi \in [1, \nu]\}. \tag{21}$$

Proof. The proof is carried out by construction. Define $T_P(s)$ to be the transfer matrix of the plant (1) with proportional feedback K_P; then

$$T_P(s)^{-1} = T(s)^{-1} - K_P \tag{22}$$

so as $\|K_P\|$ becomes sufficiently large, $T_P(0)$ approaches a diagonal matrix and $\det \overline{T}[i_1, \ldots, i_\phi](0) = 1$, which satisfies the conditions in Lemma 2.

Apply now the derivative controller in (19) so that the feedback control is described by $K_{PD}(s)$ in (20) and the closed loop transfer function is given by:

$$T_{PD}(s) = (I - T(s)K_{PD}(s))^{-1}T(s)K_{PD}(s). \tag{23}$$

Let $K_{PD}[i_1, \ldots, i_\phi](s)$ be obtained from $K_{PD}(s)$ with only the $i_1, \ldots, i_\phi$th non-redundant rows and columns. The characteristic polynomial of the nominal closed-loop system is given by:

$$d^\nu(s) - d^{\nu-1}(s) \sum_{i \in [1,\nu]} N[i]K_{PD}[i](s) + \cdots + (-1)^\nu \det(NK_{PD})(s) \tag{24}$$

whereas the characteristic polynomial for the failed system is given by:

$$d^\phi(s) - d^{\phi-1}(s) \sum_{i \in [i_1, \ldots, i_\phi]} N[i]K_{PD}[i](s)$$

$$+ d^{\phi-2}(s) \sum_{i,j \in [i_1, \ldots, i_\phi]} \det(N[i,j]K_{PD}[i,j](s))$$

$$+ \cdots + (-1)^\phi \det(N[i_1, \ldots, i_\phi](s)K_{PD}[i_1, \ldots, i_\phi](s)). \tag{25}$$

Now since $\deg K_{PD}(s) = r$, the maximum value of pole-zero difference, it follows that $\det N(s)K_{PD}(s)$ and $\det(N[i_1, \ldots, i_\phi](s)K_{PD}[i_1, \ldots, i_\phi](s))$ have the same degrees as $d^\nu(s)$ and $d^\phi(s)$ respectively. Therefore, in both cases, as $\|K_P\| \to \infty$, the polynomials are dominated by the last term of (24) and (25) respectively which are always stable from the assumption that $\det N[\phi](s), \phi = 1, 2, \ldots, \nu$, do not possess unstable zeros. □

Remark 3. In comparison with the traditional PID control, a higher derivative control D^r control is applied in this design. The practical implication of using higher derivative term is to let the closed loop system to become diagonal dominant as the frequency increases to a high value.

Given that "r", the maximum pole-zero excess, is generically 1, the PID^r controller reduces to a regular PID type where the derivative action can be indirectly synthesized as following equations (26) and (27).

In the event that high order D^r action is required, a descriptor type controller with the following structure may be used:

$$T_i^c \dot{\eta}_i = A_i^c \eta_i + B_i^c y_i, \quad i = 1, \ldots, \nu, \tag{26}$$

$$u_i = C_i^c \eta_i + D_i^c y_i, \quad i = 1, \ldots, \nu. \tag{27}$$

Where it is assumed that:

$$\text{rank}(T_i^c) - \deg(\det(sT_i^c - A_i^c)) \geq r, \quad i = 1, \ldots, \nu.$$

The following algorithm provides a procedure to synthesize a PID^r controller for plant satisfying the conditions of Theorem 2:

Algorithm 1 *Synthesis of PID^r Controller:*

1. *Verify that the conditions in Theorem 2 are satisfied.*
2. *Determine r, the maximum pole-zero excess.*
3. *Apply the proportional control (16) to the plant so that all principle minors of $\overline{T}_p(0)$ are strictly positive. This can be achieved if $\|K_P\|$ is large enough.*
4. *Synthesize a stable PD^r control (20) so that the closed loop system are asymptotically stable for all $\phi \in [1, \nu]$. This can always be achieved if the conditions in Theorem 2 are satisfied and the gain of controllers are high enough.*
5. *Apply the integral control (17) sequentially.*

3.2   Examples

Example 1 A 3-input/output system described by (1) has the the following (C, A, B) matrices and transfer matrix:

$$A = \begin{bmatrix} -3 & -3 & -1 \\ 1 & 0 & 0 \\ 0 & 1 & 0 \end{bmatrix}, \quad B = \begin{bmatrix} 0 & 0 & 1 \\ 0 & 1 & 0 \\ 1 & 0.1 & 0 \end{bmatrix}, \quad C = \begin{bmatrix} 1 & 1 & 1 \\ 0 & 1 & 1 \\ 0 & 0 & 1 \end{bmatrix}$$

$$T(s) = \frac{\begin{bmatrix} s^2 + 2s + 2 & 1.1s^2 + 1.2s + 2.2 & s^2 + s + 1 \\ s^2 + 3s + 2 & 1.1s^2 + 4.3s + 3.2 & s + 1 \\ s^2 + 3s + 3 & 0.1s^2 + 1.3s + 3.3 & 1 \end{bmatrix}}{s^3 + 3s^2 + 3s + 1}.$$

The DC gain matrices are given as:

$$T(0) = \begin{bmatrix} 2 & 2.2 & 1 \\ 2 & 3.2 & 1 \\ 3 & 3.3 & 1 \end{bmatrix}; \quad \overline{T}(0) = \begin{bmatrix} 1 & 0.6875 & 1 \\ 1 & 1 & 1 \\ 1.5 & 1.0312 & 1 \end{bmatrix}.$$

Now since

$$\det(\overline{T}[1, 3](0)) = -0.5 < 0$$
$$\det(\overline{T}[2, 3](0)) = -0.0312 < 0$$
$$\det(\overline{T}(0)) = -0.0312 < 0,$$

conditions of Lemma 2 are violated and therefore DRSPwCR for this system cannot be solved by controller (9) only. However, it is noted that all the principal minors are minimum phase:

$$\det N[1](s) = s^2 + 2s + 2$$
$$\det N[2](s) = 1.1s^2 + 4.3s + 3.2$$
$$\det N[3](s) = 1$$
$$\det N[1, 2](s) = 2s^3 + 6s^2 + 6s + 2 = 2(s + 1)^3$$

$$\det N[1,3](s) = -s^4 - 4s^3 - 6s^2 - 4s - 1 = -(s+1)^4$$
$$\det N[2,3](s) = -0.1s^3 - 0.3s^2 - 0.3s - 0.1 = -0.1(s+1)^3$$
$$\det N(s) = -s^6 - 6s^5 - 15s^4 - 20s^3 - 15s^2 - 6s - 1$$
$$= -(s+1.0039)(s+1.0019 \pm 0.0034j)(s+0.9980 \pm 0.0034j)$$
$$\times (s+0.9961).$$

Therefore, DRSPwCR can be achieved for this system by PID^r control with $r = 3$ here. In simulation, the following PID^3 is applied:

$$K_1(s) = K_2(s) = K_3(s) = -50 - 100s - 100s^2 - 50s^3 - 20/s. \quad (28)$$

The closed loop system disturbance rejection characteristic under the following failure modes are shown in Figs. 1–5.

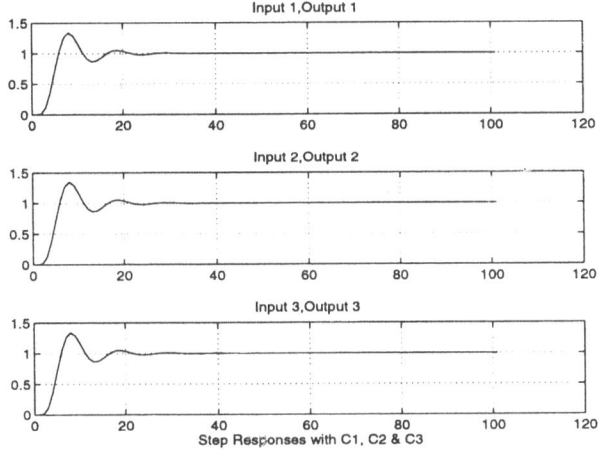

Fig. 1. System outputs at normal operation.

- All 3 channels are operational. Figure 1 shows that asymptotic tracking takes place.
- One channel has failed. Figure 2 shows the system outputs with integral control only, it is noted that the failed system remains asymptotically stable with tracking occurring for the remaining operational input/output channels. Figure 3 shows the system outputs with PID^3 control; the failed system also remains asymptotically stable with tracking occurring for the remaining operational input/output channels.
- Two of the three channels have failed. Figure 4 shows the system outputs with integral control only; this time, the fault free channel output will not be asymptotically stable. However, with PID^3 control, similar to the single controller failure scenario, asymptotic stability and regulation still hold for the remaining system as shown in Fig. 5.

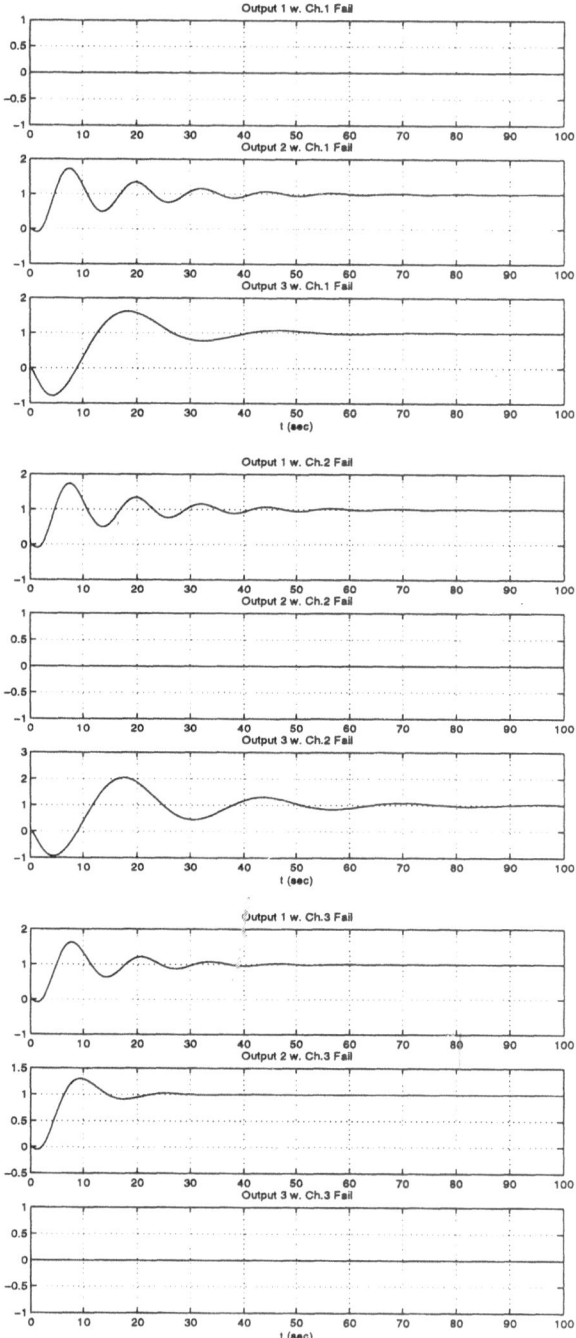

Fig. 2. Outputs with I-control, 1 channel failed.

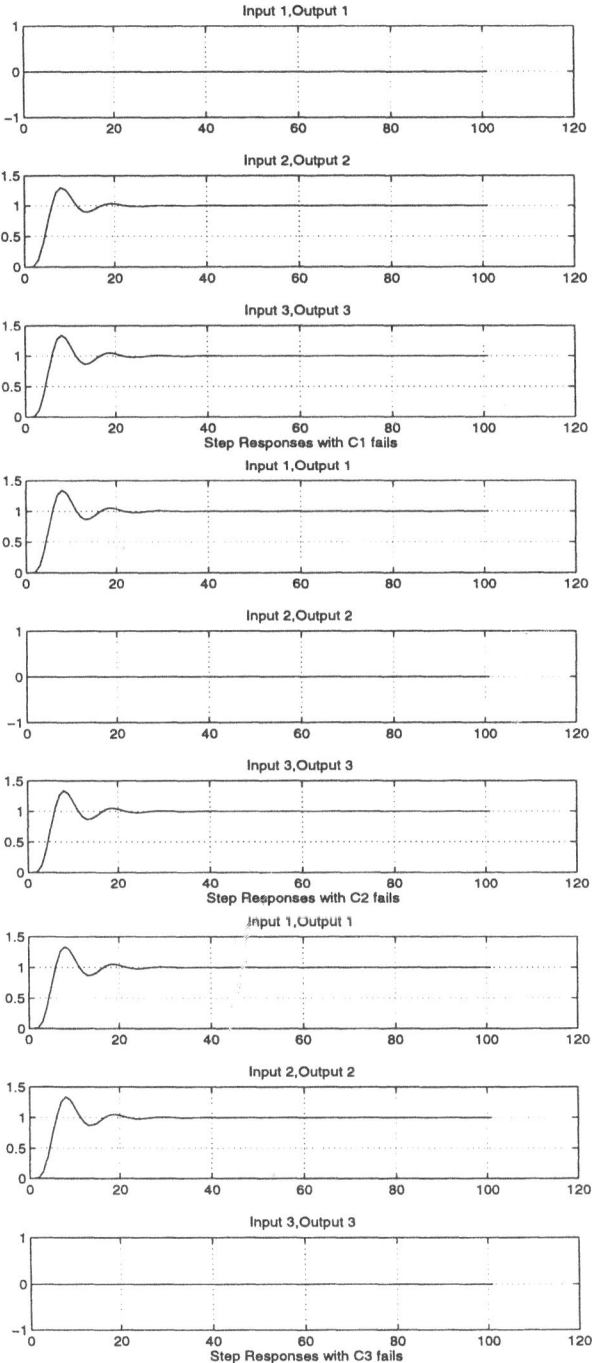

Fig. 3. Outputs with PID³ control, 1 channel failed.

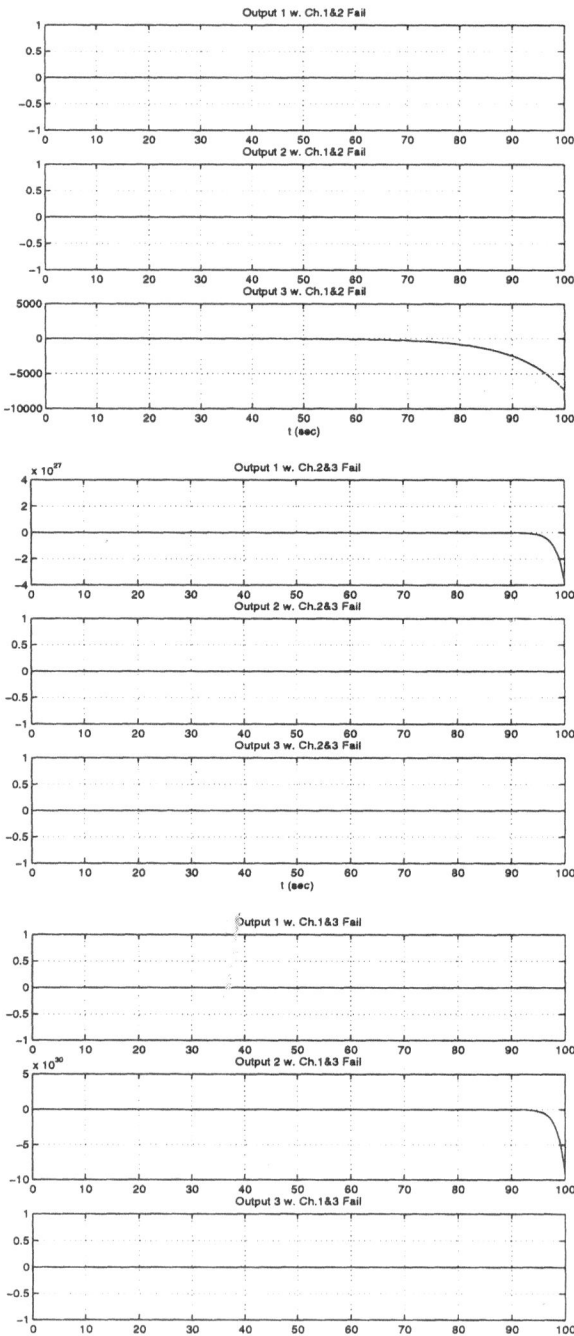

Fig. 4. Outputs with I-control, 2 channels failed.

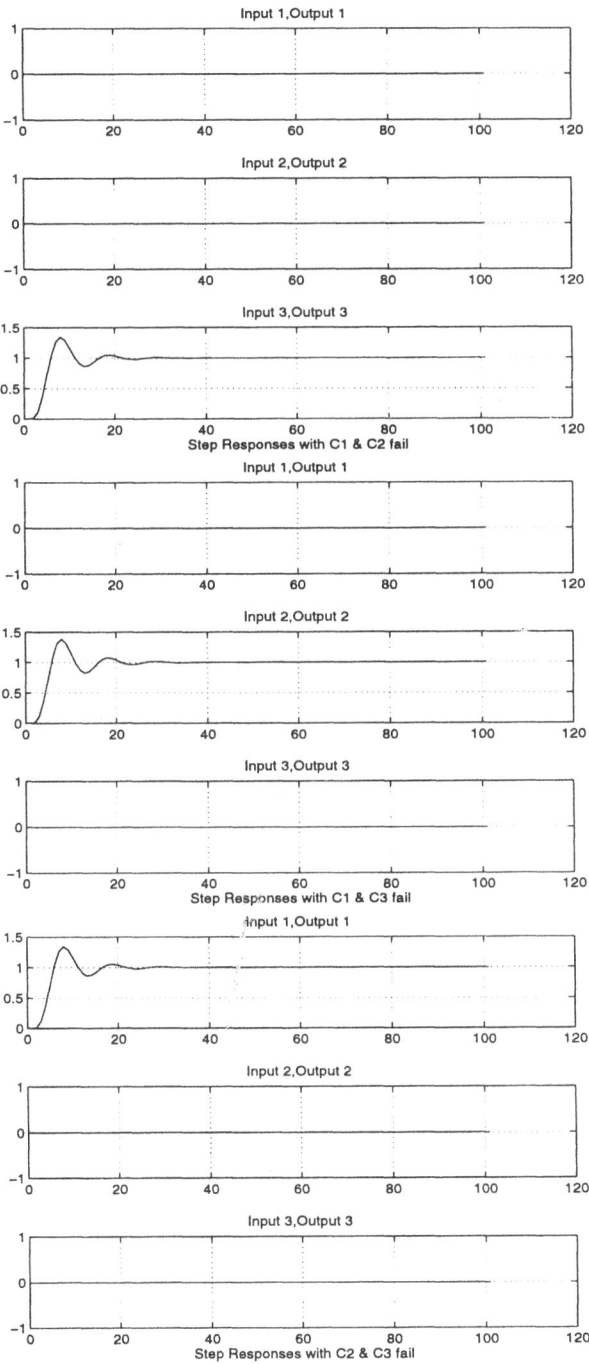

Fig. 5. Outputs with PID³ control, 2 channels failed.

Example 2 Assume now that the system parameters are perturbed; let (C', A', B') matrices be the disturbed matrices of (C, A, B) matrices, they are described as:

$$A' = \begin{bmatrix} -3.0069 & -3.0096 & -1.0299 \\ 0.9799 & -0.0123 & -0.0274 \\ 0.0162 & 1.0282 & -0.0089 \end{bmatrix}, \quad B' = \begin{bmatrix} 0.0533 & 0.0277 & 1.0122 \\ 0.0185 & 0.3781 & 0.0281 \\ 1.1458 & 0.1053 & 0.0227 \end{bmatrix}$$

$$C' = \begin{bmatrix} 1.0127 & 0.9949 & 1.2354 \\ -0.0105 & 0.8918 & 0.9919 \\ -0.0160 & -0.0075 & 1.1267 \end{bmatrix}$$

$$T(s) = \frac{\begin{bmatrix} 1.48s^2 + 3.12s + 3.18 & 0.53s^2 + 0.77s + 1.33 & 1.08s^2 + 1.12s + 1.43 \\ 1.15s^2 + 3.53s + 2.38 & 0.44s^2 + 1.75s + 1.39 & 0.03s^2 + 1.07s + 1.15 \\ 1.29s^2 + 3.93s + 3.98 & 0.11s^2 + 0.80s + 1.68 & 0.00s^2 + 0.12s + 1.32 \end{bmatrix}}{1.00s^3 + 3.02s^2 + 3.05s + 1.14}.$$

The DC gain matrices are given as:

$$T(0) = \begin{bmatrix} 3.1816 & 1.3389 & 1.4339 \\ 2.3879 & 1.3973 & 1.1502 \\ 3.9870 & 1.6891 & 1.3219 \end{bmatrix}; \quad \overline{T}(0) = \begin{bmatrix} 1 & 0.9582 & 1.0847 \\ 0.7505 & 1 & 0.8701 \\ 1.2531 & 1.2088 & 1 \end{bmatrix}.$$

Now since

$$\det(\overline{T}[2, 3](0)) = -0.0518 < 0$$
$$\det(\overline{T}(0)) = -0.1014 < 0,$$

the conditions of **Lemma 2** are violated, hence the perturbed system is still an unreliable system and all the principal minors are still minimum phase:

$$\det N[1](s) = 1.4879s^2 + 3.1242s + 3.1816$$
$$\det N[2](s) = 0.4413s^2 + 1.7530s + 1.3973$$
$$\det N[3](s) = 0.0092s^2 + 0.1215s + 1.3219$$
$$\det N[1, 2](s) = 0.0409s^4 + 1.2120s^3 + 3.4194s + s^2 + 3.3734s + 1.2485$$
$$= 0.0409(s + 26.5729)(s + 1.4985)(s + 0.7646 \pm 0.4256j)$$
$$\det N[1, 3](s) = -1.3808s^4 - 5.4979s^3 - 8.2093s^2 - 5.6110s - 1.5112$$
$$= -1.3808(s + 1.4994)(s + 0.9531)(s + 0.7646 \pm 0.4257j)$$
$$\det N[2, 3](s) = -0.0002s^4 - 0.0838s^3 - 0.2527s^2 - 0.2548s - 0.0957$$
$$= -0.0002(s + 427.59)(s + 1.5)(s + 0.76 \pm 0.43j)$$
$$\det N(s) = -0.4523s^6 - 2.7390s^5 - 6.9127s^4 - 9.4127s^3 - 7.3713s^2$$
$$- 3.1739s - 0.5957$$
$$= -0.4523(s + 1.4997)(s + 1.4993)(s + 0.7645 \pm 0.4257j)$$
$$\times (s + 0.7641 \pm 0.4256).$$

Therefore, the DRSPwCR can be achieved for this system by the same controller as (28). However, it is not necessary to use a high order derivative control since the maximum pole-zero excess is only 1 in this case, so the PID^r control with $r = 1$ is able to solve the DRSPwCR here. In simulation, the following PID^1 is applied:

$$K_1(s) = K_2(s) = K_3(s) = -50 - 100s - 20/s. \tag{29}$$

The closed loop system disturbance rejection characteristic under the following failure modes are shown in Figs. 6–8.

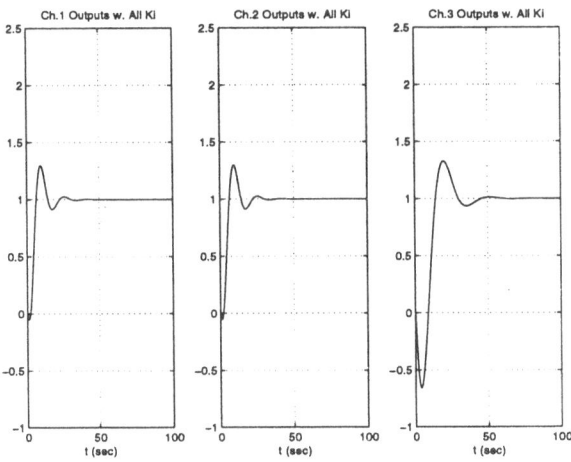

Fig. 6. System outputs at normal operation.

- All 3 channels are operational. Figure 6 shows that asymptotic tracking takes place.
- One channel has failed. Figure 7 shows the system outputs with PID^1 control, the failed system remains asymptotically stable with tracking occurring for the remaining operational input/output channels.
- Two of the three channels have failed. With PID^1 control, asymptotic stability and regulation still hold for the remaining system as shown in Fig. 8.

Summary on Simulation Results The current result shows that complete reliability against sensors and actuators failure can be achieved for the class of open loop stable, minimum phase plant which may not satisfy certain previous known conditions for reliable control. The key towards establishing reliability is the introduction of high derivative control which always stabilizes a minimum phase system without altering its DC gain. The synthesis of reliable control (for an unreliable plant) becomes a fairly straight forward process.

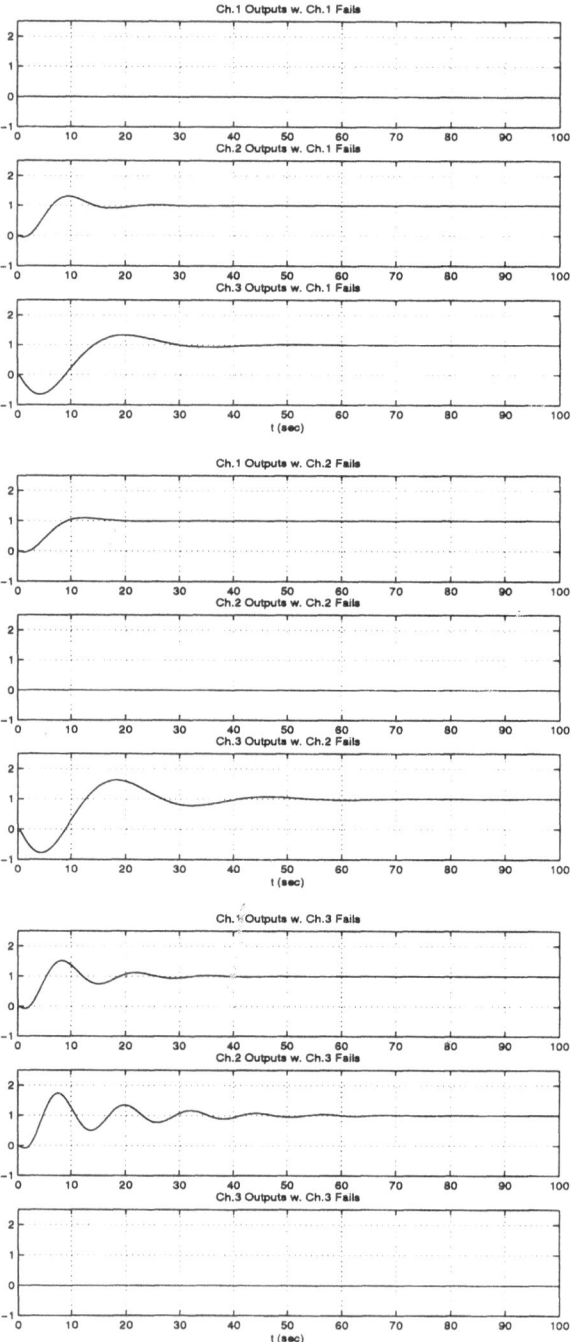

Fig. 7. Outputs with PID[1] control, 1 channel failed.

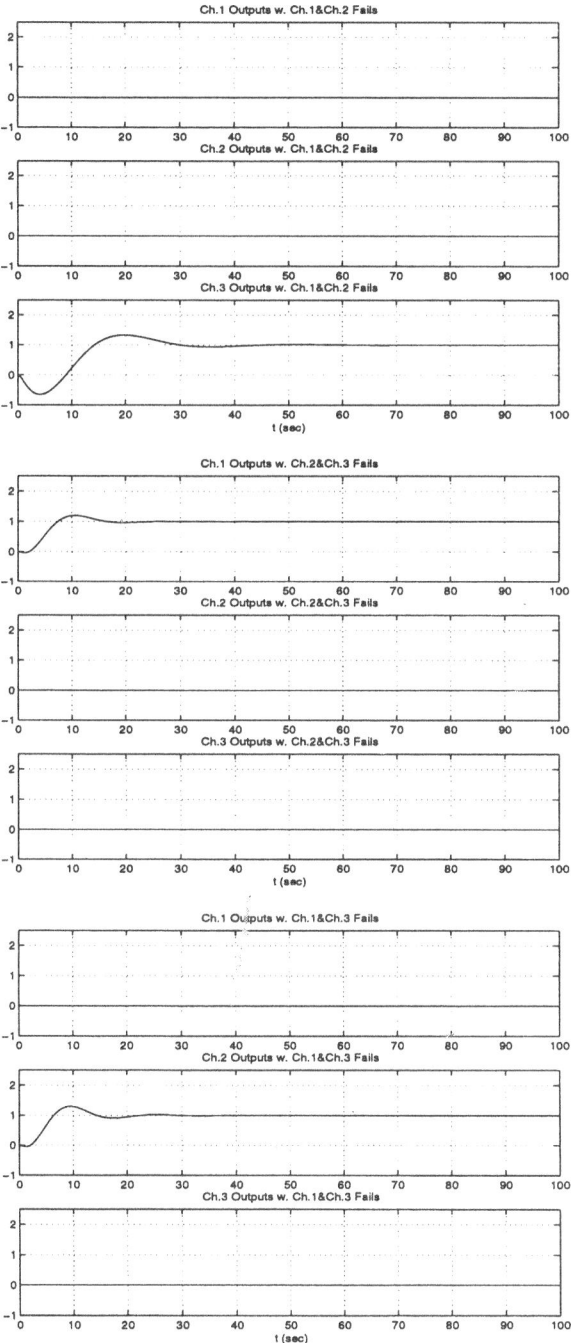

Fig. 8. Outputs with PID[1] control, 2 channels failed.

4   Block Decentralized Reliable Control Synthesis

In this section, a block decentralized control configuration is considered. That is, the feedback gain matrices possess a block diagonal structure instead of the strict diagonal structure assumed in the previous section.

Motivations for adopting a block decentralization configuration include:

1. Physical constraints such as when a number of multivariable, control subsystems are interconnected.
2. For non-minimum phase, unreliable systems, block decentralization offers an alternative way of achieving fault tolerance by grouping together strongly coupled input-output channels.

4.1   System Description

The plant is assumed to be represented by (1). To reflect the block decentralized structure, the B, C matrices are partitioned as: $B = [B_1 \ B_2 \ \cdots \ B_\mu]$, $B_i \in \mathbb{R}^{n \times n_i}$ for $i = 1, 2, \ldots, \mu$, $C = [C_1' \ C_2' \ \cdots \ C_\mu']'$, $C_i \in \mathbb{R}^{n_i \times n}$ for $i = 1, 2, \ldots, \mu$.

Correspondingly, the inputs vector v and output vector y are partitioned as:

$$u = [v_1' \ v_2' \ \cdots \ v_\mu']', v_i \in \mathbb{R}^{n_i}, \forall i = 1, 2, \ldots, \mu$$
$$y = [z_1' \ z_2' \ \cdots \ z_\mu']', z_i \in \mathbb{R}^{n_i}, \forall i = 1, 2, \ldots, \mu$$

where $\sum_{i=1}^{\mu} n_i = \nu$.

The open loop transfer matrix $T(s)$ defined in (2) is now partitioned according to the block decentralized control structure as:

$$T(s) = \begin{bmatrix} T_{11}^B(s) & T_{12}^B(s) & \cdots & T_{1\mu}^B(s) \\ \vdots & \vdots & & \vdots \\ T_{\mu 1}^B(s) & T_{\mu 2}^B(s) & \cdots & T_{\mu \mu}^B(s) \end{bmatrix} \tag{30}$$

where $T_{ij}^B \in \mathbb{C}^{n_i \times n_i}$ is the transfer submatrix corresponding to the j-th input vector u_j and i-th output vector z_i; furthermore, let

$$T_{ij}^B(s) = \frac{N_{ij}^B(s)}{d(s)}$$

where $d(s)$ is the minimal polynomial of $T(s)$ and $N_{ij}^B(s)$ is the corresponding numerator matrix. The above partition (30) corresponds to the following input-output blocks pairing:

$$\{(v_1, z_1) \ (v_2, z_2) \ \cdots \ (v_\mu, z_\mu)\} \tag{31}$$

and the feedback control is therefore block diagonal, given by:

$$U(s) = K(s)Y(s) \tag{32}$$

where

$$K(s) = \text{block diagonal}(K_1^B(s), K_2^B(s), \ldots, K_\mu^B(s)) \qquad (33)$$

and $K_i^B(s) \in \mathbb{C}^{n_i \times n_i}$.

The following block failure cases are now defined:

Definition 6. Sensor Block Failure: The i-th sensor block is said to have failed at time $t_1 > 0$ if

$$z_i(t) = 0 \quad \forall t > t_1.$$

Remark 4. Similar to the definition in the strict decentralized control structure, the i-th block sensors failure reflects the situation where all sensors in i-th block ceased to function and generate only a null output for all time thereafter.

Definition 7. Controller/Actuator Block Failure: The i-th controller and/or actuator block is said to have failed at time $t_2 > 0$ if

$$v_i(t) = 0 \quad \forall t > t_2.$$

Definition 8. System with Partial Block Failure: Assume that sensor, actuator and/or controller blocks $\{i_{\phi+1}, i_{\phi+2}, \ldots, i_\mu\} \subset \{1, 2, \ldots, \mu\}$ have failed, then the resultant system is referred to as a system with a partial block failure and it is described by the following equations:

$$\dot{x} - Ax + B_{i_1}v_{i_1} + B_{i_2}v_{i_2} + \cdots + B_{i_\phi}v_{i_\phi} + E\omega$$
$$z_i = C_i x, \quad i \in \{i_1, i_2, \ldots, i_\phi\}. \qquad (34)$$

Definition 9. ϕ-th Order Leading Principal Minor: Let $T^B[i_1, i_2, \ldots, i_\phi](s)$ be the transfer matrix of the system with a partial block failure of (34), and let

$$T^B[i_1, i_2, \ldots, i_\phi](s) = \frac{N^B[i_1, i_2, \ldots, i_\phi](s)}{d(s)}$$

when $d(s)$ is the minimal polynomial of $T(s)$ and $N^B[i_1, i_2, \ldots, i_\phi](s)$ is the corresponding numerator matrix. Then $\det(N^B[i_1, \ldots, i_\phi](s))$ is called the ϕ-th order leading principal minor.

Definition 10. Block Decentralized Robust Servomechanism Problem (BDRSP): Given the plant (1) and blocks of inputs/outputs pairing (31), obtain a block decentralized structured controller so that the following conditions all hold:

1. The closed loop system is asymptotically stable.
2. Asymptotically tracking occur, i.e. $\lim_{t\to\infty} y(t) = 0$, for all constant disturbance ω.
3. Property 2) holds for parametric perturbations: $A \to A+\delta A, B \to B+\delta B$, and $C \to C + \delta C$ provided that the closed loop system remains stable.

Definition 11. Block Transmission Zeros: Let

$$\Delta N^B[\phi] = \{\lambda \in \mathbb{C}| \det N^B[i_1, i_2, \ldots, i_\phi](\lambda) = 0,$$
$$\forall\{i_1, i_2, \ldots, i_\phi\} \subset \{1, 2, \ldots, \mu\}\}$$

be defined as the block transmission zeros of $N^B[i_1, i_2, \ldots, i_\phi](s)$.

The existing necessary and sufficient condition for BDRSP is stated as Lemma 3:

Lemma 3. [14] *There exists a solution to the BDRSP iff there exists an installation sequence $\{i_1, i_2, \ldots, i_\mu\} = \{1, 2, \ldots, \mu\}$ so that*

$$0 \notin \Delta N^B[\phi], \quad \phi = 1, 2, \ldots, \mu.$$

4.2   Main Results

Definition 12. BDRSP with Complete Reliability (BDRSPwCR): Given the open loop stable plant (1) and the block of inputs/outputs pair (31), obtain a block decentralized controller so that the following conditions all hold:

1. There exists a solution to the BDRSP for the nominal plant(1).
2. Under partial block failure, the controller solves the BDRSP for the plant (34) without retuning.

Definition 13. Steady-state Interaction Matrices: Given the plant (1) with a block decentralized control configuration, assume that the controller block installation sequence $\{i_1, i_2, \ldots, i_\mu\}$ is to be applied; then the following $\mu - 1$ steady-state interaction matrices M_i, $i = 2, \ldots, \mu$, with respect to this controller installation sequence are defined:

$$M_2[i_1, i_2] = I - T_{i_2 i_1}^B(0)(T_{i_1 i_1}^B(0))^{-1}T_{i_1 i_2}^B(0)(T_{i_2 i_2}^B(0))^{-1} \tag{35}$$

$$M_3[i_1, i_2, i_3] = I - [T_{i_3 i_1}^B(0)(T_{i_1 i_1}^B(0))^{-1} \quad T_{i_3 i_2}^B(0)(T_{i_2 i_2}^B(0))^{-1}]$$
$$\times \begin{bmatrix} I & T_{i_1 i_2}^B(0)(T_{i_2 i_2}^B(0))^{-1} \\ T_{i_2 i_1}^B(0)(T_{i_1 i_1}^B(0))^{-1} & I \end{bmatrix}^{-1}$$
$$\times \begin{bmatrix} T_{i_1 i_3}^B(0) \\ T_{i_2 i_3}^B(0) \end{bmatrix} (T_{i_3 i_3}^B(0))^{-1} \tag{36}$$

$$\vdots$$

$$M_\mu[i_1,\ldots,i_\mu] = I - [T^B_{i_\mu i_1}(0)(T^B_{i_1 i_1}(0))^{-1} \ \ldots\ T^B_{i_\mu,i_{\mu-1}}(0)(T^B_{i_{\mu-1},i_{\mu-1}}(0))^{-1}]$$

$$\times
\begin{bmatrix}
I & \cdots & T^B_{i_1,i_\mu-1}(0)(T^B_{i_\mu-1,i_\mu-1}(0))^{-1} \\
\vdots & & \vdots \\
T^B_{i_\mu-1,i_1}(0)(T^B_{i_1 i_1}(0))^{-1} \cdots & & I
\end{bmatrix}^{-1}$$

$$\times
\begin{bmatrix}
T^B_{i_1 i_\mu}(0) \\
T^B_{i_2 i_\mu}(0) \\
\vdots \\
T^B_{i_\mu-1,i_\mu}(0)
\end{bmatrix}
(T^B_{i_\mu i_\mu}(0))^{-1}
\tag{37}$$

$$\forall\{i_1,i_2,\ldots,i_\mu\} \subset \{1,2,\ldots,\mu\}.\tag{38}$$

It is noted that M_R is the Schur complement of $N^B[i_1,i_2,\ldots,i_R](0)$. Assume now the following controller is to be used:

$$\dot\xi = \Lambda^i_0\xi_i + \Lambda^i_1 v_i + \Lambda^i_2 z_i\tag{39}$$

$$\dot\eta_i = z_i\tag{40}$$

$$v_i = K^{B_i}_1 z_i + K^{B_i}_2 \xi_i + K^{B_i}\eta_i\tag{41}$$

where $i = 1,2,\ldots,\nu$ and $\xi_i \in \mathbb{R}^{S_i}$. The following results are obtained:

Theorem 3. *There exists a solution to the BDRSPwCR by using block decentralized control (39)–(41) if the following conditions all hold:*

- $0 \notin \Delta N^B[\phi], \phi = 1,2,\ldots,\mu$
- $\mathrm{Re}(\sigma(M_2[i_1,i_2])) > 0, \forall\{i_1,i_2\} \subset \{1,2,\ldots,\mu\}, i_1 \neq i_2$
- $\mathrm{Re}(\sigma(M_3[i_1,i_2,i_3])) > 0, \forall\{i_1,i_2,i_3\} \subset \{1,2,\ldots,\mu\}, i_1 \neq i_2 \neq i_3$
- $\vdots$
- $\mathrm{Re}(\sigma(M_\mu[i_1,\ldots,i_\mu])) > 0, \forall\{i_1,\ldots,i_\mu\} \subset \{1,\ldots,\mu\}, i_1 \neq i_2 \neq \cdots \neq i_\mu$

where $\mathrm{Re}(\sigma(M(\cdot)))$ are the real parts of eigenvalues of the matrices $M(\cdot)$.

Remark 5. In general, if the plant model (1) of the plant is not available, the steady-state interaction matrices defined above can be obtained experimentally.

To solve the BDRSPwCR, it is not always adequate to use only proper controllers. That is, the conditions of Theorem 3 are not always satisfied by any general process. Therefore, a block PID^r type controller, which is the multivariable extension of (16)–(20), will be applied to weaken the conditions of Theorem 3. The structure of the block PID^r controller is given below:

$$K_P^B(s) = \text{block diag}(K_P^1, K_P^2, \ldots, K_P^\mu) \tag{42}$$

$$K_I^B(s) = \text{block diag}\left(\frac{K_I^1}{s}, \frac{K_I^2}{s}, \ldots, \frac{K_I^\mu}{s}\right) \tag{43}$$

$$K_{Di}^B(s) = K_{di}^1 s + K_{di}^2 s^2 + \cdots + K_{di}^r s^r, i = 1, 2, \ldots, \mu \tag{44}$$

$$K_D^B(s) = \text{block diag}(K_D^1(s), K_D^2(s), \ldots, K_D^\mu(s)) \tag{45}$$

$$K_{PD}^B(s) = K_P(s) + K_D(s) \tag{46}$$

$$K^B(s) = K_P(s) + K_I(s) + K_D(s) \tag{47}$$

where

$$r = \text{order}(d(s)) - \min\{\Delta N^B[1], \Delta N^B[2], \ldots, \Delta N^B[\mu]\}. \tag{48}$$

Theorem 4. *There exists a solution to the BDRSPwCR (Definition 12) if the following conditions all hold:*

1. *There is a solution to BDRSP for (1).*
2. *$\Delta N^B[\phi] < 0$, $\phi = 1, 2, \ldots, \mu$.*

Proof. Define $T_P(s)$ to be the transfer matrix of plant (1) with block decentralized proportional feedback $K_P(s) = \text{block diag}(K_P^1, K_P^2, \ldots, K_P^\mu)$; then,

$$T_P(s)^{-1} = T(s)^{-1} - K_P(s) \tag{49}$$

as $\|K_P^i\| \to \infty, \forall i \in [1, \nu]$, $T_P(0)$ becomes diagonal and matrices $M_2[i_1, i_2], \ldots,$ $M_\mu[i_1, i_2, \ldots, i_\mu] \to I$, so that $\text{Re}(\sigma(M_i)) > 0$, thus satisfying the conditions in Theorem 3.

For the consideration of stability under partial groups of channel failure, assume that $T^B[i_1, \ldots, i_\phi](s)$ is the subsystem consists of the fault free blocks:

$$T^B[i_1, \ldots, i_\phi](s) = \frac{1}{d(s)} N^B[i_1, \ldots, i_\phi](s) \tag{50}$$

and let $K_{PD}^B[i_1, i_2, \ldots, i_\phi](s)$ be the corresponding ϕ-th order fault-free block decentralized PD^r controller applying to the system with partial block failure of (34). The fault-free subsystem closed-loop transfer matrix is then given by:

$$
\begin{aligned}
T_{PD}^B[i_1, \ldots, i_\phi](s) &= (I - T^B[i_1, \ldots, i_\phi](s)K_{PD}^B[i_1, \ldots, i_\phi](s))^{-1} T^B[i_1, \ldots, i_\phi] \\
&= (d(s)I - N^B[i_1, \ldots, i_\phi]K_{PD}^B[i_1, \ldots, i_\phi])^{-1} N^B[i_1, \ldots, i_\phi].
\end{aligned}
$$

Now set $\|K_P\| \to \infty, \|K_D\| \to \infty$, and let r be the maximum pole-zero excess of $T(s)$:

$$
\begin{aligned}
&T_{PD}^B[i_1, \ldots, i_\phi](s) \\
&\to -(K_{PD}^B[i_1, \ldots, i_\phi](s)^{-1} N^B[i_1, \ldots, i_\phi](s)^{-1}) N^B[i_1, \ldots, i_\phi](s) \tag{51}
\end{aligned}
$$

when the determinant of $N^B[i_1, \ldots, i_\phi](s) \neq 0$ $\forall s \in [0, +\infty]$, i.e., $\Delta N^B[\phi] < 0$, $\phi = 1, \ldots, \mu$, then

$$N^B[i_1, \ldots, i_\phi](s)^{-1} N^B[i_1, \ldots, i_\phi](s) = I.$$

Therefore equation (51) becomes:

$$T_{PD}^B[i_1, \ldots, i_\phi](s) \to -K_{PD}^B[i_1, \ldots, i_\phi](s)^{-1}.$$

The controller $K_{PD}^B[i_1, \ldots, i_\phi](s)$ matrix can always be chosen such that each controller is a polynomial with all roots being located in LHP. Therefore, the roots of the characteristic polynomial of the closed loop system are located in LHP. □

Remark 6. Specially, for a two-block system, condition 2 of Theorem 4 becomes

$$\mathrm{Re}(\Delta N^B[1]) < 0 \tag{52}$$
$$\mathrm{Re}(\Delta N^B[2]) < 0, \tag{53}$$

i.e. $\det N^B[1](s), \det N^B[2](s)$ are Hurwitz.

The following example in the next section is a 4-input/output system with its transfer matrix being partitioned into two blocks and the simulation results confirm the above theorem, i.e., the BDRSPwCR can be solved by using the block decentralized PID^r control.

4.3   Example

In this example, the BDRSPwCR will be solved for a 4-input/output system consisting of 2 blocks with 2 input/output channels. The principal diagonal blocks $T_{11}^B(s)$ and $T_{22}^B(s)$ are modified Rosenbrock's Models (Rosenbrock's Model with the 2 inputs interchanged). The system does not satisfy the conditions in Theorem 3; however, it satisfies the conditions in Theorem 4, so the BDRSPwCR can be solved by block PID^r controllers.

The transfer function matrix is partitioned into two blocks, with each block consists a 2-input/output subsystem:

$$T_{11}^B(s) = \frac{1}{d(s)} N_{11}^B(s) = \frac{\begin{bmatrix} -s^2 + s + 2 & -s^2 + 1 \\ -s^2 + 1 & -s^2 - 0.6667s + 0.3333 \end{bmatrix}}{s^3 + 3s^2 + 3s + 1} \tag{54}$$

$$T_{12}^B(s) = \frac{1}{d(s)} N_{12}^B(s) = \frac{\begin{bmatrix} 100s^2 + 100s + 100 & 10s^2 + 10s + 20 \\ 10s^2 + 10s + 30 & 10s^2 + 10s + 10 \end{bmatrix}}{s^3 + 3s^2 + 3s + 1} \tag{55}$$

$$T_{21}^B(s) = \frac{1}{d(s)} N_{21}^B(s) = \frac{\begin{bmatrix} 100s^2 + 100s + 100 & 10s^2 + 10s + 20 \\ 10s^2 + 10s + 30 & 10s^2 + 10s + 10 \end{bmatrix}}{s^3 + 3s^2 + 3s + 1} \tag{56}$$

$$T_{22}^B(s) = \frac{1}{d(s)} N_{22}^B(s) = \frac{\begin{bmatrix} -s^2 + s + 2 & -s^2 + 1 \\ -s^2 + 1 & -s^2 - 0.6667s + 0.3333 \end{bmatrix}}{s^3 + 3s^2 + 3s + 1}. \quad (57)$$

The DC gain sub-matrices are

$$T_{11}^B = \begin{bmatrix} 2 & 1 \\ 1 & 0.3333 \end{bmatrix} \quad (58)$$

$$T_{12}^B = \begin{bmatrix} 100 & 20 \\ 30 & 10 \end{bmatrix} \quad (59)$$

$$T_{21}^B = \begin{bmatrix} 100 & 20 \\ 30 & 10 \end{bmatrix} \quad (60)$$

$$T_{22}^B = \begin{bmatrix} 2 & 1 \\ 1 & 0.3333 \end{bmatrix}. \quad (61)$$

It is noted that for the $T_{11}^B(s), T_{22}^B(s)$ blocks, reliable control for strict diagonal decentralized configuration cannot be obtained as discussed in the previous chapter.

The steady-state interaction matrix is

$$\begin{aligned} M_2 &= I - T_{21}^B(0)(T_{11}^B(0))^{-1}T_{12}^B(0)(T_{22}^B(0))^{-1} \\ &= \begin{bmatrix} 1.599 \times 10^3 & 1.8 \times 10^3 \\ 0 & -8.99 \times 10^2 \end{bmatrix}. \end{aligned} \quad (62)$$

The eigenvalues of the M_2 matrix are calculated as -1.599×10^3, -8.99×10^2, which are all negative, and the condition in Theorem 3 is not satisfied. However, the transmission zeros of all principal minor blocks are given by:

$$\begin{aligned} \text{roots}(\det N_{11}^B(s)) &= -1, -1, -1 \\ \text{roots}(\det N_{22}^B(s)) &= -1, -1, -1 \\ \text{roots}(\det N(s)) &= -0.52491 \pm j1.0989, \\ &\quad -0.47891 \pm j0.99090, \\ &\quad -0.49546 \pm j0.29998, \\ &\quad -0.50408 \pm j0.27644. \end{aligned}$$

The real parts of all the block transmission zeros are all negative, so the conditions in Theorem 4 are satisfied and the BDRSPwCR can be solved by the block decentralized PID^r structured controllers. In the example, the maximum pole-zero excess is 1, hence the PID^1 controllers are selected as:

$$K_1(s) = \begin{bmatrix} -50s - 50 - 1/s & 0 \\ 0 & -50s - 50 - 1/s \end{bmatrix} \quad (63)$$

$$K_2(s) = \begin{bmatrix} -50s - 50 - 1/s & 0 \\ 0 & -50s - 50 - 1/s \end{bmatrix}. \quad (64)$$

4.4   Simulation Results

The following simulations are made for two cases: (1) Without PD[1] feedback controllers and (2) With PD[1] controller enhancement. These situations are considered:

- Both blocks are normally operational.
- Block 2 has failed.
- Block 1 has failed.

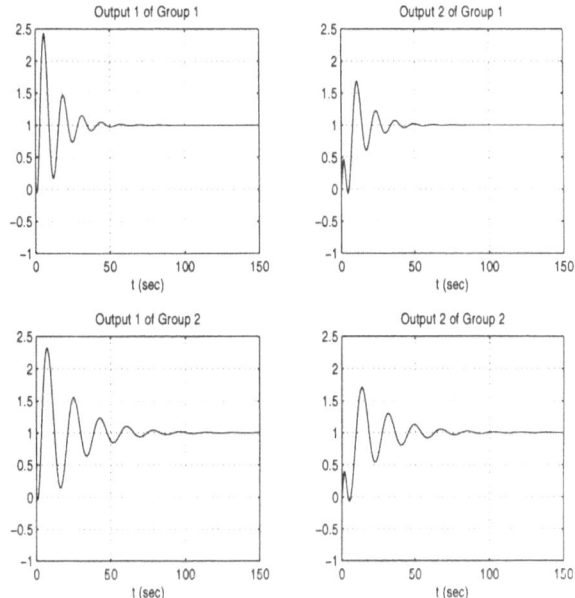

Fig. 9. Step responses of the plant at normal operation.

Case 1: Without PD[1] feedback control: Block decentralized integral controller with

$$K_I^1(s) = K_I^2(s) = \begin{bmatrix} 1/s & 0 \\ 0 & 1/s \end{bmatrix}$$

are applied to the plant. As shown in Fig. 9, asymptotic regulation takes place for the nominal plant. However, the closed loop system becomes explosively unstable when controller 1 fails (Fig. 10) or controller 2 fails (Fig. 11).

Case 2: With PD[1] feedback control: As shown in Figs. 12, 13 and 14, with the block decentralized PID[1] controllers (63) and (64) added, the closed loop system is reliable. The failure of any block does not affect the stability of the system and the fault-free block continues to produce asymptotic regulation.

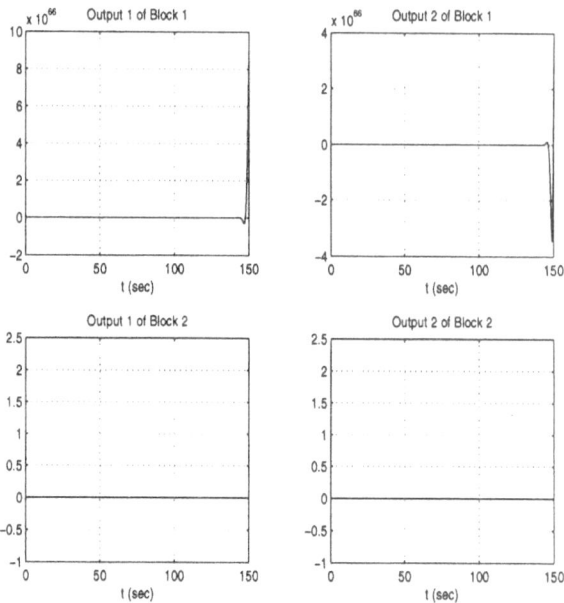

Fig. 10. Step responses of the plant with block 2 failed.

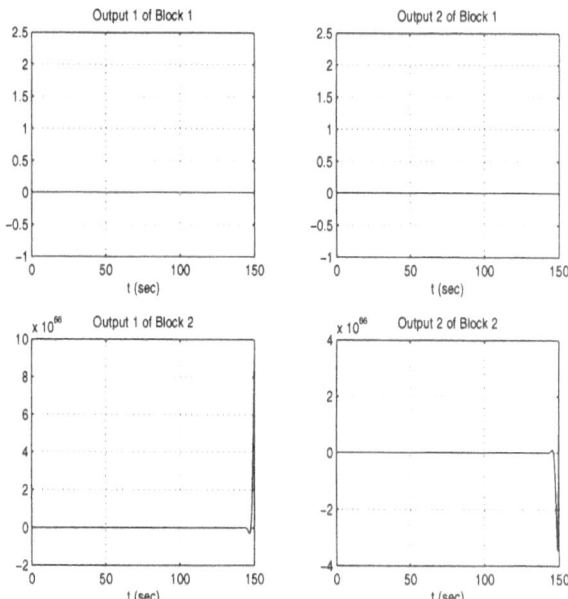

Fig. 11. Step responses of the plant with block 1 failed.

It is concluded that for non-minimum phase, unreliable, open-loop stable linear systems, it is necessary to group the strongly coupled channels, and treat this group as a single sub-system. It should be noted that the responses of the closed loop system have not been optimized. In the event that faster speed of response is desired, the parameter optimization method [3] can be applied.

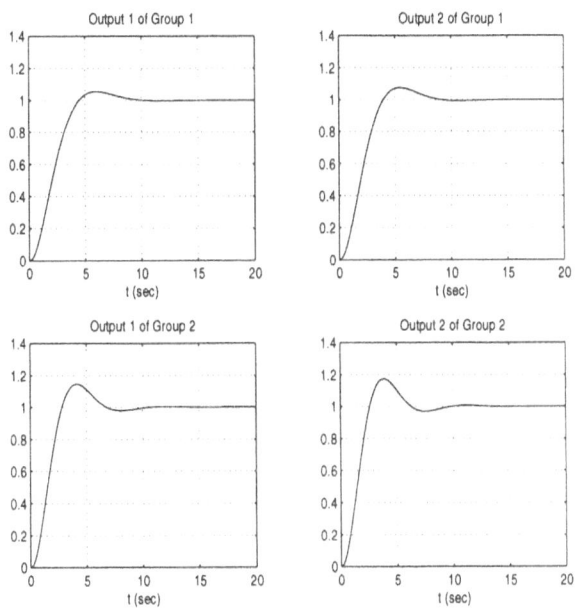

Fig. 12. Step responses of the plant at normal operation.

5   Conclusions

In this paper, a decentralized reliable control system for large-scale, multi-input/output linear plant has been developed. First, DRSPwCR is solved for the class of plants that are open loop stable, unreliable, and minimum phase by applying the strict decentralized PID^r algorithm. This is followed by considering the BDRSPwCR which is solved for the plants with non-minimum phase minors where block diagonal decentralized controllers are applied. The following conclusions are obtained:

- Reliable control can be achieved for a class of open loop linear system with the steady-state interaction indices satisfying certain conditions by using decentralized integral controller configurations.
- By applying the decentralized proportional feedback controllers, the steady-state interaction indices can be adjusted to satisfy the reliable control

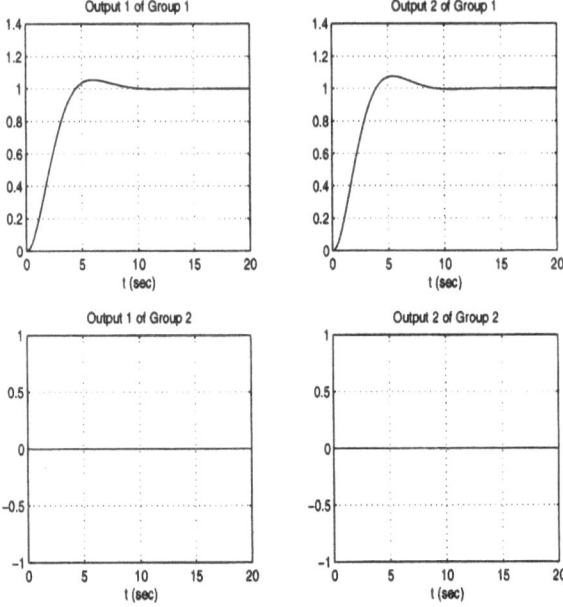

Fig. 13. Response of the plant with block 2 failed.

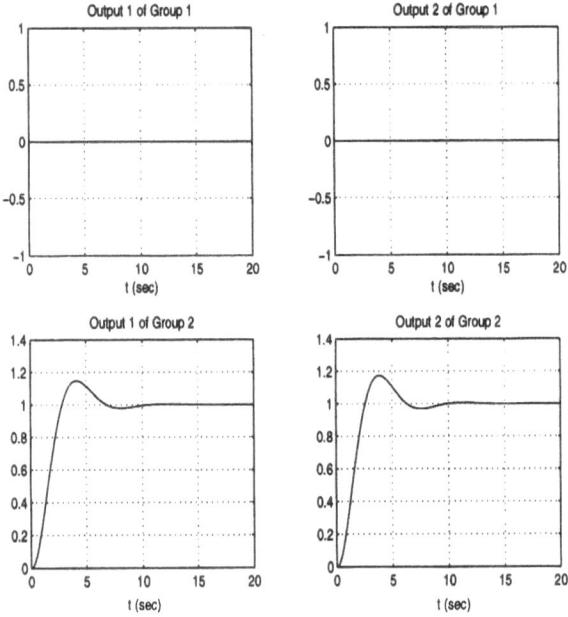

Fig. 14. Response of the plant with block 1 failed.

conditions but will destabilize the closed loop system under certain input-output channels failure.

- Introduction of a decentralized D^r controller to re-stabilize the closed loop system does not affect the steady-state interaction indices. Therefore, the synthesis of decentralized PID^r controllers can be carried out in the following sequence: First, apply proportional control to alter the steady state interaction indices. Second, synthesize D^r control to re-stabilize the closed loop system. Third, apply integral control to provide tracking and disturbance rejection.

- The block diagonal decentralized control configuration is an alternative way to solve the reliable control problem for the systems which have non-minimum phase minors and cannot achieve reliable control by strict diagonal decentralized control.

6   Appendix: Necessary Condition of Reliability for 2 I/O Systems

In this appendix, the necessary conditions to achieve reliable control for a given unreliable plant are analyzed inductively. The general controller synthesis procedures are derived, followed by 2 numerical examples of reliable control using different control strategies.

In order to develop the theoretical conditions for reliable control by using strict diagonal decentralized control, it is first necessary to analyze the range of reliability of proportional controller K_P for the general 2×2 open loop stable system.

Given an open loop stable, unreliable plant of the structure

$$T(s) = \frac{1}{d(s)} \begin{bmatrix} n_{11}(s) & n_{12}(s) \\ n_{21}(s) & n_{22}(s) \end{bmatrix} \tag{65}$$

where $d(s)$ is a stable polynomial. Let

$$d(s) = s^n + d_{n-1}s^{n-1} + \cdots + d_1 s + d_0 \tag{66}$$

$$n_{ij}(s) = n_{r_{ij}}^{ij} s^{r_{ij}} + n_{r_{ij}-1}^{ij} s^{r_{ij}-1} + \cdots + n_1^{ij} s + n_0^{ij} \tag{67}$$

where $i, j \in \{1, 2\}$ and r_{ij} denote the order of the polynomial $n_{ij}(s)$. To streamline the proof, it is assume that

$$n_{ij}(0) > 0, \quad i, j \in \{1, 2\}.$$

while the proof for the other cases can be carried out in a similar manner. Define now

$$\Delta n(s) = n_{11}(s)n_{22}(s) - n_{12}(s)n_{21}(s)$$

and assume that a decentralized proportional controller

$$K_P = \begin{bmatrix} k_{p1} & 0 \\ 0 & k_{p2} \end{bmatrix}$$

has been applied. The regions of reliability for the proportional gain parameters are now analyzed as follows:

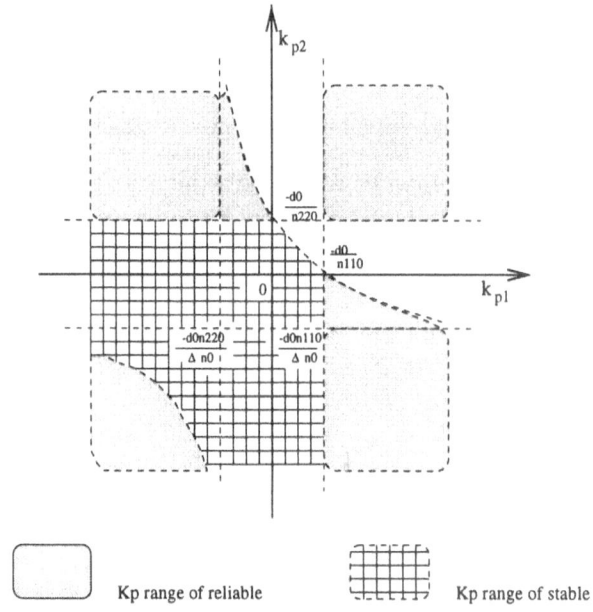

Fig. 15. K_P range of stability and reliability.

As shown in Fig. 15, in order for the closed loop system to be reliable, the values of k_{p1} and k_{p2} have to be located in the shaded areas described by the following four equations:

Region 1 (Quadrant I):

$$k_{p1} > \frac{d_0}{n_{11}(0)}, \quad k_{p2} > \frac{d_0}{n_{22}(0)}. \tag{68}$$

Region 2 (Quadrant II):

$$\left(\frac{d_0 n_{22}(0)}{\Delta n(0)} - k_{p1} \right) \left(\frac{d_0 n_{11}(0)}{\Delta n(0)} - k_{p2} \right) < \frac{d_0^2 n_{12}(0) n_{21}(0)}{\Delta n(0)^2}$$

$$k_{p1} > \frac{d_0}{n_{11}(0)}, \quad k_{p2} < 0. \tag{69}$$

Region 3 (Quadrant III):

$$\left(\frac{d_0 n_{22}(0)}{\Delta n(0)} - k_{p1} \right) \left(\frac{d_0 n_{11}(0)}{\Delta n(0)} - k_{p2} \right) < \frac{d_0^2 n_{12}(0) n_{21}(0)}{\Delta n(0)^2}$$

$$k_{p1} < 0, \quad k_{p2} > \frac{d_0}{n_{22}(0)}. \tag{70}$$

Region 4 (Quadrant IV):

$$\left(\frac{d_0 n_{22}(0)}{\Delta n(0)} - k_{p1}\right)\left(\frac{d_0 n_{11}(0)}{\Delta n(0)} - k_{p2}\right) > \frac{d_0^2 n_{12}(0)n_{21}(0)}{\Delta n(0)^2}$$
$$k_{p1} < 0, \quad k_{p2} < 0. \tag{71}$$

The derivation of the reliable range of K_P can be found in Appendix C. When decentralized PD^r controllers are applied, the closed loop characteristic polynomials are described by the following equations:

$$D_1(s) = d(s) - n_{11}(s)k_{p1} \tag{72}$$
$$D_2(s) = d(s) - n_{22}(s)k_{p2} \tag{73}$$
$$D_{12}(s) = D_1(s)D_2(s) - n_{12}(s)n_{21}(s)k_{p1}k_{p2} \tag{74}$$

where $D_1(s)$ is the closed loop system characteristic polynomial when k_{p1} only is applied, similarly $D_2(s)$ is the closed loop system characteristic polynomial when k_{p2} only is applied and $D_{12}(s)$ is the closed loop characteristic polynomial when both proportional controllers are applied.

Define $\overline{T}_P$ the normalized closed loop transfer matrix as described in (22). T_1 is the closed loop DC gain matrix when controller 1 only is installed, while T_2 is the closed loop DC gain matrix when controller 2 only is installed. For the reliable control consideration, values of k_{p1}, k_{p2} have to be such that the closed loop system $\overline{T}_P$ possesses positive principal minors under both cases of controllers normal operations and any channel failures; therefore, the conditions in one of the following scenarios must be satisfied:

Scenario 1:

$$\det(\overline{T}_P) > 0 \tag{75}$$
$$T_P(1,1)T_1(1,1) > 0 \tag{76}$$
$$T_P(2,2)T_2(2,2) > 0. \tag{77}$$

Scenario 2:

$$\det(\overline{T}_P) < 0 \tag{78}$$
$$T_P(1,1)T_1(1,1) < 0 \tag{79}$$
$$T_P(2,2)T_2(2,2) > 0. \tag{80}$$

Scenario 3:

$$\det(\overline{T}_P) < 0 \tag{81}$$
$$T_P(1,1)T_1(1,1) > 0 \tag{82}$$
$$T_P(2,2)T_2(2,2) < 0. \tag{83}$$

T_P, T_1 and T_2 are related to the values of open loop DC gain matrix and K_P values as described by the following equations:

$$T_1(1,1) = \frac{n_{11}(0)}{D_1(0)} \tag{84}$$

$$T_2(2,2) = \frac{n_{22}(0)}{D_2(0)} \tag{85}$$

$$T_P(1,1) = \frac{d_0 n_{11}(0) - \Delta n(0) k_{p2}}{D_{12}(0)} \tag{86}$$

$$T_P(2,2) = \frac{d_0 n_{22}(0) - \Delta n(0) k_{p1}}{D_{12}(0)} \tag{87}$$

$$\det(\overline{T}_P) = \frac{D_{12}(0)}{(d_0 n_{11}(0) - \Delta n(0) k_{p2})(d_0 n_{22}(0) - \Delta n(0) k_{p1})} D_{12}(0). \tag{88}$$

Furthermore,

$$D_1(0) = d_0 - n_{11}(0) k_{p1} \tag{89}$$

$$D_2(0) = d_0 - n_{22}(0) k_{p2} \tag{90}$$

$$D_{12}(0) = D_1(0) D_2(0) - n_{12}(0) n_{21}(0) k_{p1} k_{p2}. \tag{91}$$

Substituting (84), (85), (86), (87), and (88) into the above 3 scenarios, and multiplying all three equations of any one of the scenarios together, the following equation is obtained as a necessary requirement of K_P for reliability consideration:

$$D_1(0) D_2(0) D_{12}(0) < 0. \tag{92}$$

Therefore, in order to achieve reliable control, K_P values have to be such that:

1. All closed loop characteristic polynomials in equations(11), (12) and (13) must be stable.
2. K_P values must be located in the ranges described in (68), (69), (70) and (71) of Fig. 15.

References

1. T. N. Chang and E. J. Davison, "Interaction indices for the decentralized control of unknown multivariable systems," in *Proceedings to the 10th IFAC world Congress*, Munich, West Germany, vol.8, pp. 257-262, 1987.
2. T. N. Chang and E. J. Davison, "Reliability of robust decentralized controllers," in *Proceedings to the 4th IFAC/IFORS Symposium on Large-scale Systems: Theory and Applications*, Zurich, Switzerland, pp. 297-392, 1986.
3. E. J. Davison and T. N. Chang, "Decentralized controller design using parameter optimization methods," *Control Theory and Advanced Technology*, vol. 2, pp. 131-154, 1986.

4. Z. F. Chen and T. N. Chang, "Synthesis of reliable control systems," in *Proceedings of The American Control Conference*, Albuquerque, New Mexico, pp. 3473-3474, 1997.

5. C. A. Desoer and and A. N. Gündeş, "Stability under sensor or actuator failures", *Recent Advances in Robust Control*, P. Dorato and R. K. Yedavalli, editors, IEEE Press, pp. 185-186, 1990.

6. A. N. Gündeş, M. G. Kabuli, "Reliable decentralized control," in *Proceedings of the American Control Conference*, Baltimore, Maryland, pp. 3359-3363, 1994.

7. A. N. Gündeş, "Stabilizing controller design for linear system with sensor or actuator failures," *IEEE Transactions on Automatic Control*, vol.39, pp. 1224-1230, 1994.

8. A. N. Gündeş, "Stability of feedback systems with sensor or actuator failures: analysis," *International Journal of Control*, vol.56, pp. 735-753. 1992.

9. R. J. Veillettea, J. V. Medanić and W. R. Perkins, "Design of reliable control systems," *IEEE Transactions on Automatic Control*, vol.37, pp. 290-304, 1992.

10. M. H. Shor and W. J. Kolodziej, "Reliable control to actuator signal attenuation-type faults," in *Proceedings of the 32nd Conference on Decision and Control*, San Antonio, Texas, pp. 3418-3419, 1993.

11. M. H. Shor and W. R. Perkins, "Decentralized control with A prescribed degree of stability: a unified discrete/continuous-time design," *Proceedings of the American Control Conference*, San Francisco, California, pp. 2396-2397, 1993.

12. R. J. Veillette, J. V. Medanić and W. R. Perkins, "Robust stabilization and disturbance rejection for uncertain systems by decentralized control," in *Proceedings of the Workshop on Control of Uncertain Systems: Progress in System and Control Theory*, D. Hinrichsen and B. Martensson, Eds., Cambridge, Massachusetts: Birkhauser, pp. 309-327, 1989.

13. R. J. Veillette, J. V. Medanić and W. R. Perkins, "Robust stabilization and disturbance rejection for systems with structured uncertainty," in *Proceedings of 28th IEEE Conference on Decision and Control*, Tampa, Florida, pp. 936-941, 1989.

14. E. J. Davison, "Decentralized robust control of unknown systems using tuning regulators," *IEEE Transactions on Automatic Control*, AC-23, pp. 276-289, 1978.

A Novel Approach to Vibration Reduction in Flexible Belt Drives*

Robert B. Gorbet and Scott A. Bortoff

Department of Electrical and Computer Engineering, University of Toronto,
Toronto, Ontario, Canada M5S 3G4
Email: rbgorbet, bortoff@control.toronto.edu

Abstract. This paper presents a novel method to compensate the effects of flexibility in belt-driven drive trains. Such drives are found in robot manipulators, for example, where drive flexibility is undesirable. Unfortunately, the effects of flexibility can be difficult to compensate using the primary actuator, due to bandwidth limitations. By introducing a second actuator, which applies a force to the belt at a point between the pulleys, we show that the closed-loop dynamics can be made to be more "stiff" and even completely rigid. A backstepping design and a linear LQR design are outlined, and simulations of the latter illustrate improved performance over classical control which neglects the effects of flexibility.

1 Introduction

Flexible materials are often used to couple actuators to mechanical systems. For example, the links of a robotic manipulator may be driven by a flexible belt or cable, because this allows the servos to be mounted at the stationary base, reducing manipulator mass and inertia. If the flexible dynamics are neglected in the design of a position or force controller, closed-loop performance may suffer, and instability can result. This observation has motivated much research into the compensation of flexible drive train dynamics [1, e.g.]. For the control system designer, this remains a challenging problem for two reasons. First, the flexible members make the system under-actuated, meaning there is less than one actuator per degree of freedom. Second, the system is often singularly-perturbed, meaning the flexible members are stiff and therefore exhibit higher-frequency behavior relative to the rigid part of the system.

To understand the challenges, consider the system illustrated in Fig. 1. J_i is the pulley inertia, q_i is the pulley angle and r_i is the pulley radius for both the output pulley (unactuated, $i = 1$), and the input pulley (actuated, $i = 2$). If the belt has infinite stiffness, and we assume no slip at the pulleys, the pulley angles are related by the algebraic constraint

$$r_1 q_1 = r_2 q_2. \tag{1}$$

* This research supported by the Natural Sciences and Engineering Research Council of Canada (NSERC) under grant OGP0138423, and the NSERC Postdoctoral Fellowship Program. Please direct correspondence to S. A. Bortoff, phone: (416)-978-0562, fax: (416)-978-0804

The system has one degree of freedom and one control input, and the dynamics are

$$\frac{r_1^2 J_2 + r_2^2 J_1}{r_1 r_2}\ddot{q}_1 = u_2. \tag{2}$$

States (q_1, q_2) which satisfy (1) are said to be on the *rigid manifold*. When the belt has infinite stiffness, state evolution is constrained to this manifold.

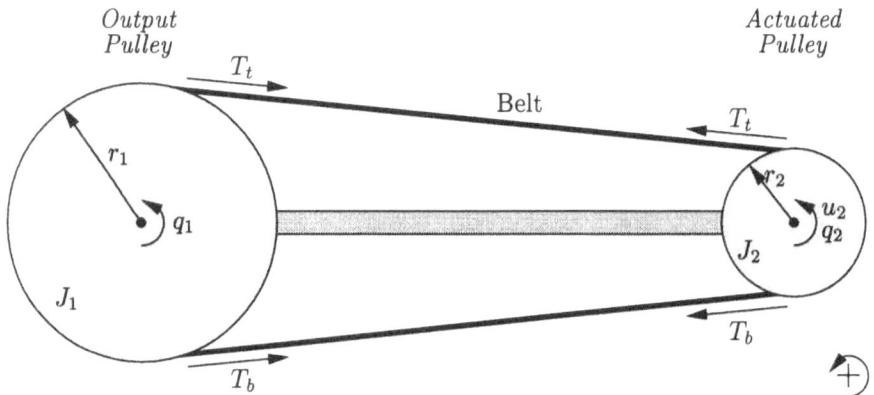

Fig. 1. Belt-driven system

When the belt spring constant, k, is finite, this constraint no longer exists. The belt can stretch, altering the tensions in the top and bottom spans, T_t and T_b. When $r_1 q_1 > r_2 q_2$, for example, T_t increases by $k(r_1 q_1 - r_2 q_2)$ and T_b decreases by the same amount. Since the nominal tension is the same in both spans, the equations of motion are (neglecting damping terms)

$$J_1\ddot{q}_1 = r_1(T_b - T_t) \qquad = \quad 2kr_1(r_2 q_2 - r_1 q_1), \tag{3}$$
$$J_2\ddot{q}_2 = r_2(T_t - T_b) + u_2 = -2kr_2(r_2 q_2 - r_1 q_1) + u_2. \tag{4}$$

The system now has two degrees of freedom, but only one control input u_2. It is under-actuated.

Let us now assume the system is stiff, and write the dynamics in singularly perturbed form. Define the constants

$$\bar{J} = \frac{r_1^2 J_2 + r_2^2 J_1}{r_1 r_2}, \qquad J^* = \frac{J_1 J_2}{r_1^2 J_2 + r_2^2 J_1}, \qquad \text{and} \quad \varepsilon = \sqrt{\frac{J^*}{2k}},$$

and the coordinate transformation

$$x_1 = q_1, \qquad \eta_1 = \frac{2kr_1}{J_1}(r_2 q_2 - r_1 q_1),$$

$$x_2 = \dot{q}_1, \qquad \eta_2 = \varepsilon\frac{2kr_1}{J_1}(r_2 \dot{q}_2 - r_1 \dot{q}_1).$$

In these new coordinates, the dynamics (3) and (4) are

$$\dot{x}_1 = x_2, \tag{5}$$
$$\dot{x}_2 = \eta_1, \tag{6}$$
$$\varepsilon \dot{\eta}_1 = \eta_2, \tag{7}$$
$$\varepsilon \dot{\eta}_2 = -\eta_1 + u_2/\bar{J}. \tag{8}$$

If k is large relative to J_1 and J_2, then $\varepsilon \ll 1$, and there is significant time-scale separation between the "fast" η-subsystem and the "slow" x-subsystem. As $k \to \infty$, $\varepsilon \to 0$ and the slow system results:

$$\dot{\bar{x}}_1 = \bar{x}_2, \tag{9}$$
$$\dot{\bar{x}}_2 = u_2/\bar{J}. \tag{10}$$

Thus, $\bar{J}$ is the inertia of the rigid part of the system, and with $\bar{x}_1 = q_1$, (9)–(10) are exactly the rigid dynamics (2).

"Neglecting the flexible dynamics" means setting $\varepsilon = 0$ in (5)–(8), and designing a controller for (9)–(10). The PD controller

$$u_2 = u_{\mathrm{s}} = \bar{J}[-k_p(\bar{x}_1 - x_1^d) - k_d(\bar{x}_2 - x_2^d) + \dot{x}_2^d], \tag{11}$$

where $k_p > 0$ and $k_d > 0$ are the position and velocity gains, and x_1^d, x_2^d and $\dot{x}_2^d$ are the desired position, velocity and acceleration signals, globally exponentially stabilizes the origin of the trajectory tracking error of the reduced system (9)–(10). If damping is included in the dynamics, singular perturbation theory guarantees that the control (11) will also globally exponentially stabilize the full system (5)–(8), even for some small non-zero values of ε (Tikhonov's Theorem, [2, e.g.]). This requires that k be large compared to J^*, and also that k_p and k_d not be so large as to destroy the time-scale separation.

The challenging problem to the control engineer is when ε is not small enough to guarantee stability, or gains must be increased to achieve the required performance specifications. In this case, (11) must be modified. One approach is to add a so-called "fast" component, u_{f}, to the slow control u_{s} (11), redefining u_2 as

$$u_2 = u_{\mathrm{s}} + u_{\mathrm{f}}.$$

The fast control can improve the performance of the fast (flexible) part of (5)–(8). For example, it can add damping to the fast dynamics. Or, it can effectively decrease ε, and hence increase the time-scale separation. This approach is known as composite control [3, e.g.]. On paper, composite control gives the designer two controls, u_{f} for the second-order fast dynamics, and u_{s} for the second-order slow dynamics, using only one control channel, u_2. If the frequency components of the two controls u_{s} and u_{f} are separate, they will not interfere with each other. On paper, the system is now fully actuated.

Unfortunately, composite control has a disadvantage, especially in systems where inertias J_1 and/or J_2 are large: A single physical actuator is required to generate both u_s and u_f. If the inertias are large, u_s will contain large amplitude torque commands, requiring a large actuator. At the same time, u_f will contain high-frequency, often high-amplitude spectral components. The single actuator may not be able to satisfy both of these requirements simultaneously.

This paper proposes a new approach. We introduce a second physical actuator called a *dancer*, which directly compensates the flexible dynamics in a belt-driven drive train. The idea comes from tension control in web-handling machines [4], where the position of a so-called "dancer roll" – a large cylinder in contact with the web – is adjusted in order to control web tension. While passive "idler" or "tensioner" pulleys have been in use in automotive belt drives for close to two decades [5], we believe this is the first work to consider active tension control of a drive belt.

We show that the dancer actuator is better suited than the drive actuator for the role of canceling the effect of belt flexibility, because it can have a lower inertia than either pulley, and hence a wider bandwidth. Furthermore, because the dancer actuator is an independent mechanical subsystem, and because its control is decoupled from the slow control, as we shall see, existing drive trains can be retro-fitted in order to reduce or eliminate the effects of flexibility. An increase in performance is thereby achieved without a full redesign of the original system.

The dynamics of the dancer system are introduced in the next section, followed by an analysis of its effectiveness in compensating flexible dynamics. Section 3 then summarizes two different control designs. Finally, simulations of the system are provided in Sect. 4.

2   The Dancer Actuator

Figure 2 shows the guided pulley system. The dancer is the link with the two small idler pulleys at either end. The dancer translates linearly, up and down as a unit, orthogonal to the horizontal link, and is actuated by a force u_3; the distance between the two idlers is fixed at $2d_3$. When the dancer moves up or down, it changes the relative tensions in the two belt spans. The rotational inertia of the two idlers is assumed to be zero, as is the mass of the belt, so that the belt tension is the same on the left-hand and right-hand sides of the dancer.

Let q_3 denote the dancer position, with $q_3 = 0$ denoting the center position, and J_3 denote the dancer mass. Then the dynamics of the dancer-actuated system are

$$J_1 \ddot{q}_1 = r_1(T_b - T_t), \tag{12}$$

$$J_2 \ddot{q}_2 = r_2(T_t - T_b) + u_2, \tag{13}$$

$$J_3 \ddot{q}_3 = T_t[\cos(\alpha_1) + \cos(\alpha_2)] - T_b[\cos(\beta_1) + \cos(\beta_2)] + u_3. \tag{14}$$

When compared with the system of Fig. 1, there are now two sources of belt stretching: relative pulley rotation ($r_1 q_1 \neq r_2 q_2$), and dancer motion q_3.

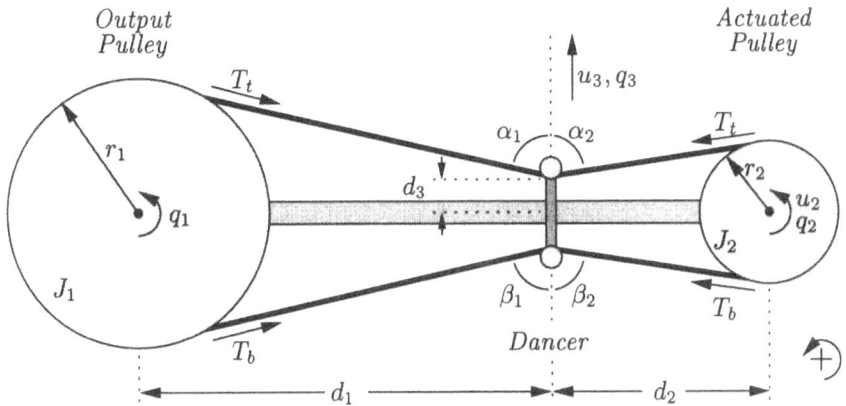

Fig. 2. Belt-driven system with dancer

In order to calculate belt stretching due to dancer motion, the kinematics of the system must be examined. Figure 3 shows a closeup view of the bottom half of pulley 1. The solid belt represents the nominal position, the dashed belt, the position for some positive q_3. For clarity, the dancer pulley is not shown.

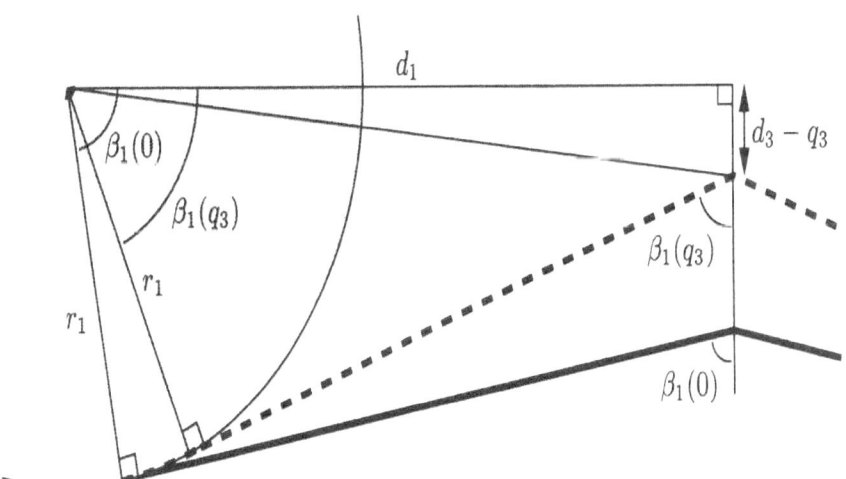

Fig. 3. Belt kinematics

$$\beta_1(q_3) = \arctan\left(\frac{d_3 - q_3}{d_1}\right) + \arctan\left(\frac{\sqrt{d_1^2 + (d_3 - q_3)^2 - r_1^2}}{r_1}\right).$$

Similarly, the other angles in Fig. 2 are:

$$\alpha_1(q_3) = \arctan\left(\frac{d_3 + q_3}{d_1}\right) + \arctan\left(\frac{\sqrt{d_1^2 + (d_3 + q_3)^2 - r_1^2}}{r_1}\right),$$

$$\alpha_2(q_3) = \arctan\left(\frac{d_3 + q_3}{d_2}\right) + \arctan\left(\frac{\sqrt{d_2^2 + (d_3 + q_3)^2 - r_2^2}}{r_2}\right),$$

$$\beta_2(q_3) = \arctan\left(\frac{d_3 - q_3}{d_2}\right) + \arctan\left(\frac{\sqrt{d_2^2 + (d_3 - q_3)^2 - r_2^2}}{r_2}\right).$$

The length of the dashed line in Fig. 3 (left of the dancer location) is

$$\sqrt{d_1^2 + (d_3 - q_3)^2 - r_1^2} + r_1[\beta_1(0) - \beta_1(q_3)];$$

that of the solid line is $\sqrt{d_1^2 + d_3^2 - r_1^2}$. From this, the *change* in length, *due to dancer motion only*, of the top and bottom belt spans of Fig. 2, l_t and l_b is:

$$l_t(q_3) = \sqrt{d_1^2 + (d_3 + q_3)^2 - r_1^2} + r_1[\alpha_1(0) - \alpha_1(q_3)] - \sqrt{d_1^2 + d_3^2 - r_1^2}$$

$$+ \sqrt{d_2^2 + (d_3 + q_3)^2 - r_2^2} + r_2[\alpha_2(0) - \alpha_2(q_3)] - \sqrt{d_2^2 + d_3^2 - r_2^2},$$

$$l_b(q_3) = \sqrt{d_1^2 + (d_3 - q_3)^2 - r_1^2} + r_1[\beta_1(0) - \beta_1(q_3)] - \sqrt{d_1^2 + d_3^2 - r_1^2}$$

$$+ \sqrt{d_2^2 + (d_3 - q_3)^2 - r_2^2} + r_2[\beta_2(0) - \beta_2(q_3)] - \sqrt{d_2^2 + d_3^2 - r_2^2}.$$

Using this notation, and assuming superposition can be used to combine the two sources of belt stretching, the tensions in the belt spans are

$$T_t = T_0 - k(r_2 q_2 - r_1 q_1) + k l_t(q_3), \tag{15}$$

$$T_b = T_0 + k(r_2 q_2 - r_1 q_1) + k l_b(q_3). \tag{16}$$

T_0 is the nominal belt tension. For clarity, we further define the non-linear functions ϕ, γ, ψ and ζ:

$$\phi(q_3) = \frac{l_b(q_3) - l_t(q_3)}{2},$$

$$\gamma(q_3) = -T_0[\cos(\alpha_1(q_3)) + \cos(\alpha_2(q_3)) - \cos(\beta_1(q_3)) - \cos(\beta_2(q_3))],$$

$$\psi(q_3) = \frac{l_b(q_3)[\cos(\beta_1(q_3)) + \cos(\beta_2(q_3))]}{2} - \frac{l_t(q_3)[\cos(\alpha_1(q_3)) + \cos(\alpha_2(q_3))]}{2},$$

$$\zeta(q_3) = \frac{\cos(\alpha_1(q_3)) + \cos(\alpha_2(q_3)) + \cos(\beta_1(q_3)) + \cos(\beta_2(q_3))}{2}.$$

Then, substituting (15) and (16) into (12)–(14) and using the above notation, the equations of motion are

$$J_1 \ddot{q}_1 = 2kr_1(r_2 q_2 - r_1 q_1) + 2kr_1 \phi(q_3), \tag{17}$$

$$J_2 \ddot{q}_2 = -2kr_2(r_2 q_2 - r_1 q_1) - 2kr_2 \phi(q_3) + u_2, \tag{18}$$

$$J_3 \ddot{q}_3 = -2k(r_2 q_2 - r_1 q_1)\zeta(q_3) - 2k\psi(q_3) - \gamma(q_3) + u_3. \tag{19}$$

Remark 1. Although they appear complicated, the functions ϕ, ψ, ζ and γ are well-defined for all values of q_3 within the valid region of operation, $|q_3| < \min\{r_1, r_2\} - d_3$. Moreover, ϕ, γ and ψ are well-approximated by their linearization, while ζ is well-approximated by $\zeta(0)$. This implies that the system (17)–(19) is well-approximated by its Jacobian linearization.

Remark 2. The functions ϕ, ψ and ζ are functions of only q_3 and kinematic parameters – not dynamic parameters such as inertias or k. Thus, they can be computed precisely. The function γ is proportional to the nominal tension T_0, which may be uncertain. Since it is in range of the control u_3, however, such uncertainty can be easily compensated. Moreover, γ is an odd function, and is "negative" stabilizing feedback.

Remark 3. The dancer location, determined by d_1 and d_2, is chosen so that the angles $\alpha_1(0) = \alpha_2(0) = \beta_1(0) = \beta_2(0)$. This considerably simplifies the dynamics.

2.1 Why Add a Second Control?

In Sect. 1, we considered the fourth-order system (3)–(4) to be a cascade consisting of a double integrator (5)–(6), which represents the rigid part of the system, driven by a fast actuator (7)–(8), which represents the flexible part. In this section, we do the same for the dancer system (17)–(19). That is, we may consider the dancer system to be a double-integrator driven in cascade by *two* actuator subsystems, each of which are second-order systems. Writing the dynamics in such a form will allow us to ascertain the benefits of using the dancer control u_3 to compensate the flexible dynamics, over using the pulley control u_2.

Again, define constants

$$\varepsilon = \sqrt{\frac{J^*}{2k}}, \quad \text{and} \quad \delta = \sqrt{\frac{J_3}{2k\phi_0'\zeta_0}}, \tag{20}$$

and states

$$x_1 = q_1, \qquad \eta_1 = \frac{2kr_1}{J_1}(r_2 q_2 - r_1 q_1), \qquad \xi_1 = \frac{2kr_1}{J_1}\phi(q_3),$$

$$x_2 = \dot{q}_1, \qquad \eta_2 = \varepsilon\frac{2kr_1}{J_1}(r_2 \dot{q}_2 - r_1 \dot{q}_1), \qquad \xi_2 = \delta\frac{2kr_1}{J_1}\phi'(q_3)\dot{q}_3.$$

The 0 subscript denotes evaluation at 0 and $(\cdot)'$ denotes differentiation with respect to q_3, e.g. $\phi_0' = \partial\phi/\partial q_3(0)$. Then the system (17)–(19) can be expressed as

$$\dot{x}_1 = x_2, \tag{21}$$

$$\dot{x}_2 = \eta_1 + \xi_1, \tag{22}$$

$$\varepsilon\dot{\eta}_1 = \eta_2, \tag{23}$$

$$\varepsilon\dot{\eta}_2 = -\eta_1 - \xi_1 + u_2/\bar{J}, \tag{24}$$

$$\delta\dot{\xi}_1 = \xi_2, \tag{25}$$

$$\delta\dot{\xi}_2 = -\frac{\phi'\zeta}{\phi_0'\zeta_0}\eta_1 - \frac{\phi'\psi}{\phi_0'\zeta_0\phi}\xi_1 - \frac{r_1\phi'}{J_1\phi_0'\zeta_0}\gamma + \frac{r_1J_3}{J_1\phi_0'\zeta_0}\phi''\dot{q}_3^2 + \frac{r_1\phi'}{J_1\phi_0'\zeta_0}u_3, \tag{26}$$

where the argument q_3 of ϕ, ψ, etc., is suppressed in (26). Written this way, it is obvious that the six-dimensional system can be considered to be a double integrator (21)–(22), driven in cascade by two "fast" second-order actuators, represented by the η-dynamics (23)–(24), and the ξ-dynamics (25)–(26), respectively.

Remark 4. The ξ-subsystem dynamics (25)–(26) are normalized, so that the Jacobian linearization is obvious by inspection. Each of the five terms on the right-hand side of (26) is well-defined, in the sense that there is no division by zero.

In these coordinates, the η-actuator dynamics (23)–(24) are linear. The bandwidth of the actuator is approximately $1/\varepsilon$, and its DC-gain is $1/\bar{J}$. On the other hand, linearizing the ξ-dynamics, (and reusing the ξ-notation) gives[1]

$$\delta\dot{\xi}_1 = \xi_2,$$

$$\delta\dot{\xi}_2 = -\eta_1 - \xi_1 - \frac{\gamma_0'}{2k\phi_0'\zeta_0}\xi_1 + \frac{r_1}{J_1\zeta_0}u_3.$$

For $k \gg 1$, the bandwidth of the ξ-actuator is approximately $1/\delta$. Notice from (20) that δ is proportional to $\sqrt{J_3}$, which is likely to be significantly smaller than the $\sqrt{J^*}$ appearing in ε. Since the constants ϕ_0' and ζ_0 are on the order of 10^{-1}, the bandwidth of the ξ-subsystem can be expected to be wider than that of the η-subsystem if in fact J_3 is small enough compared to J^*. Moreover, the DC gain of the ξ-system is

$$\frac{2k\phi_0'\zeta_0}{2k\phi_0'\zeta_0 + \gamma_0'} \cdot \frac{r_1}{J_1\zeta_0} \approx \frac{r_1}{J_1\zeta_0}$$

for $k \gg 1$, which is the same order as the DC gain for the η-system. Thus, using the ξ-actuator to compensate the high-frequency dynamics due to flexibility is well-founded on a bandwidth argument, even for very large values of k. As an example, the system parameters used in simulation in Sect. 4 are shown in Table 1. The approximate DC gain and bandwidth of each actuator is shown in Table 2.

---

[1] Recall that $(\cdot)'$ denotes differentiation with respect to q_3.

Table 1. Parameters for the simulations, in MKS units

Parameter	Value	Parameter	Value
r_1	0.2	k_p	16.0
r_2	0.1	k_d	8.0
k	50000.0	d_1	1.0
J_1	1.0	d_2	0.335
J_2	0.5	d_3	0.05
J_3	0.1	T_0	100.0

Table 2. DC gain and bandwidth for each actuator subsystem

System parameter	η-subsystem	ξ-subsystem
DC gain	0.666	0.627
Bandwidth (rad/s)	77.46	292.85

3   Control Designs

The control philosophy pursued here is to decouple, or approximately decouple the system. That is, we shall design u_2 for the rigid system, as in (11), and design u_3 to cause the system to be *exactly* rigid, which is an integral manifold approach, or "more" rigid than it is without the dancer, which is a high-gain approach. This decoupled approach allows the retro-fit of existing systems with the dancer subsystem, making them behave exactly like, or more like, a rigid system, without affecting any design which already exists for u_2.

Both of the designs considered in this section are simplified by transforming the system (17)–(19) into the following coordinates:

$$
\begin{aligned}
y_1 &= q_2, & y_4 &= \dot{q}_2, \\
y_2 &= r_2 q_2 - r_1 q_1, & y_5 &= r_2 \dot{q}_2 - r_1 \dot{q}_1, \\
y_3 &= q_3, & y_6 &= \dot{q}_3.
\end{aligned}
$$

3.1   Backstepping

Backstepping is an iterative procedure allowing the design of a control which globally exponentially stabilizes the origin of the error dynamics in a tracking problem. The key to the design is the generation, at each iteration, of a "virtual control" which guarantees the negative definiteness of a candidate Lyapunov function. The procedure results in a full coordinate transformation to the error coordinates and a real control which guarantees global exponential stability of the origin. See, for example, [2].

In this instance, backstepping can be applied to define a control law u_3 which forces the rigid manifold $r_1 q_1 = r_2 q_2$ to be attractive and invariant. Define the "output" of the system to be y_2. We seek a controller to force $y_2 \to 0$. In the process of the backstepping design, we will generate a coordinate transformation $y_i \rightarrow z_i$, and the resulting control law u_3 will guarantee that $z_i \rightarrow 0$, accomplishing the goal.

First, define $z_1 = y_2$ and the candidate Lyapunov function $V_1 = z_1^2/2$. Differentiating V_1 gives $\dot{V}_1 = z_1 y_5$. We now define the first "virtual control": If y_5 were a control input, then the control $y_5 = -c_1 z_1$, $c_1 > 0$, ensures $\dot{V}_1$ is negative definite, and $z_1 = y_2 \rightarrow 0$. At this point, the design would be complete. However, y_5 is not a control. Define instead a virtual control $v_1 = -c_1 z_1$, the error coordinate $z_2 = y_5 - v_1$, and the new candidate Lyapunov function $V_2 = V_1 + z_2^2/2$. This is the beginning of the second iteration of the design.

Differentiating V_2 now gives

$$\dot{V}_2 = z_1 y_5 + z_2 \dot{z}_2 = z_1(z_2 + v_1) + z_2 \dot{z}_2 = -c_1 z_1^2 + z_2(z_1 + \dot{z}_2). \qquad (27)$$

Substitute $\dot{z}_2 = r_2 \ddot{q}_2 - r_1 \ddot{q}_1 - \dot{v}_1$ in (27), choose another virtual control to make $\dot{V}_2$ negative definite, define another error coordinate z_3, and so on. The process iterates until an actual control input appears in $\dot{z}_i$. At this point, we stop defining error coordinates and compute an actual control to make $\dot{V}_i$ negative definite.

In the case of the dancer system, the real control u_3 appears in $\dot{z}_4$. The control which results from the backstepping design has the remarkable property that it forces the 6-dimensional system to the rigid manifold, so the input-output behavior from u_2 to q_1 is exactly second-order (rigid). The zero dynamics are fourth-order, defined by the z-dynamics that are computed in the backstepping procedure, and are globally exponentially stable. The drawback is that the control u_2, which is pre-defined by (11), is differentiated twice during the backstepping algorithm, and so the desired trajectory x_1^d, x_2^d, etc., must be sufficiently smooth. The backstepping control is also somewhat computationally complex. This design is treated in detail in [6].

3.2   LQR Linear Design

A considerably simpler approach can render the rigid manifold approximately invariant using high-gain. We simply linearize the dynamics of (17)–(19), and express them in the y-coordinates,

$$\dot{y} = Ay + B_2 u_2 + B_3 u_3. \qquad (28)$$

Assume the control u_2 is given by (11), and include this feedback in (28). Then we design u_3 to minimize the cost

$$J = \int_0^\infty y^T Q y + R u_3^2 \, dt$$

subject to $\dot{y} = Ay - B_2[k_p(y_1 - y_1^d) + k_d(y_4 - y_4^d) + \ddot{y}_4^d] + B_3 u_3$. We choose Q to put a large penalty on deviation from the rigid manifold (y_2 and y_5), and R is chosen small. This results in a very simple linear control law that does not depend explicitly on u_2.

4 Simulations

In this section, we give the results of two simulations for the system (17)–(19). The system parameters were defined in Table 1. For both simulations, the initial condition is

$$[q_1(0)\ q_2(0)\ q_3(0)\ \dot{q}_1(0)\ \dot{q}_2(0)\ \dot{q}_3(0)]^T = [0\ 0.1\ 0\ 0\ 0\ 0]^T,$$

which is off the rigid manifold and therefore excites the flexible dynamics. Then at $t = 1.0$, the desired trajectory q_1^d is a smooth unit step (a 5th-order polynomial of t), reaching $q_1^d(t) = 1.0$ for all $t \geq 1.25$. In this way, we observe both the zero-input behaviour as well as the tracking response.

In the first simulation, shown in Fig. 4, $u_3 = 0$ and u_2 is given by (11). This shows the natural response of the system. Initially, the higher tension in the lower span causes the pulleys to rotate towards each other. They then oscillate 180° out of phase, as the PD controller damps out the vibrations. The effect of the dancer oscillations is visible as a high-frequency vibration. Note that the scales for dancer displacement and pulley rotation are different.

The step response is typical of a flexible system. Note that damping has not been included in the system dynamics, so all damping comes from the PD control u_2.

For the simulation parameters used, $\varepsilon = 0.013$ and $\delta = 0.003$. The resulting time-scale separation is obvious, with the dancer oscillating at a higher frequency than the pulleys. The high- and low-frequency components of the control u_2 are clearly visible.

For the second simulation, u_3 is designed using the LQR control outlined in the previous section, with

$$Q = \text{diag}\{0, 1, 0.1, 0, 0.1, 0.01\}, \qquad R = 0.000001.$$

Solution of the LQR problem in Matlab gives the following feedback gains in the y-coordinates:

$$k_3 = [\ 9.808,\ 6094,\ 1938,\ 4.355,\ -122.8,\ 101.9\ ].$$

Note that some of these numbers seem large due to the use of MKS units, but the control effort which results is reasonable. The simulation results are shown in Fig. 5. As before, the pulleys initially rotate towards each other, but the dancer rapidly removes any oscillation in the system: By $t = 0.5$ seconds, the system is very nearly on the rigid manifold. The step response is much closer to that of a rigid system, with no oscillation and minimal lag between pulley angles. Note that both controls are smooth, compared to u_2 of Fig. 2. Also, dancer travel is limited to two centimeters in compensating for the initial condition, and less than one centimeter in tracking the step.

Figure 6 shows the state y_2, the deviation from the manifold $r_1 q_1 = r_2 q_2$, for both simulations. The LQR control improves the convergence to the rigid manifold. However, the manifold is not invariant, and control u_2 does excite the

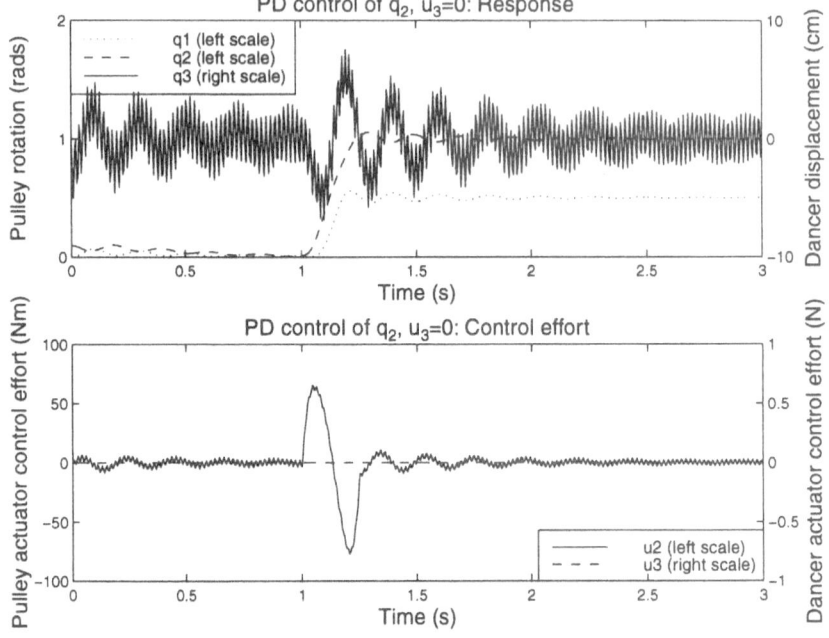

Fig. 4. System response with $u_3 = 0$

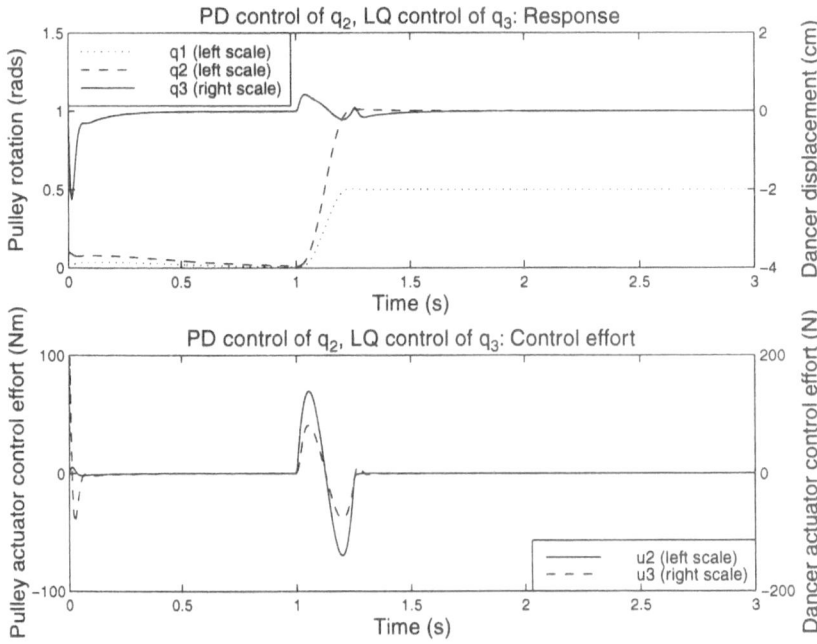

Fig. 5. System response with u_3 designed using LQR

flexible dynamics. The backstepping control of Sect. 3.1 forces $y_2 \to 0$ exactly, and makes the manifold invariant so that flexibility is not excited during the step [6].

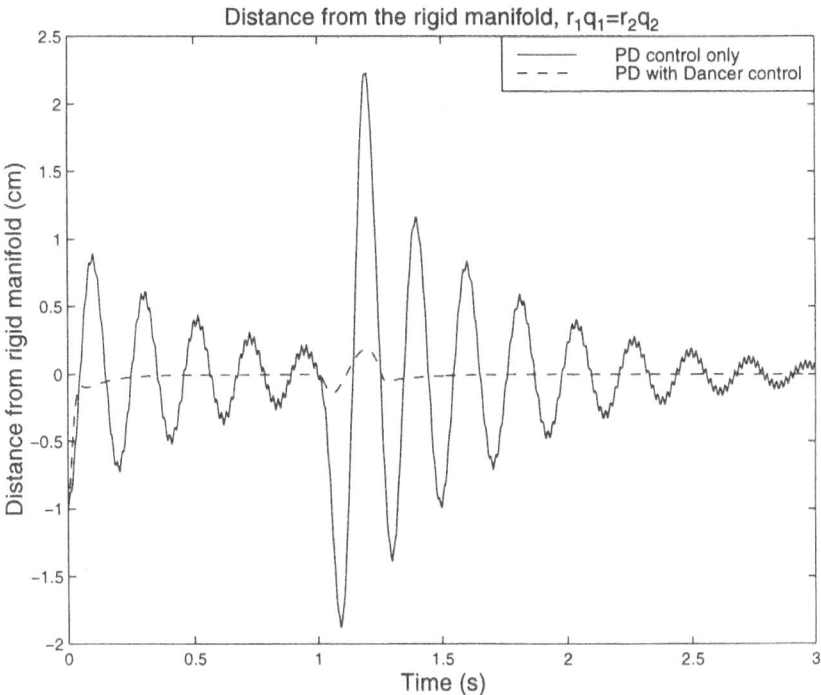

Fig. 6. Deviation from the rigid manifold for both controls

5   Conclusion

This paper presented a novel mechanical method to compensate actively the flexible effects in belt-driven drive trains. By introducing a second actuator, the flexible dynamics are easily controlled. This actuator, which we call a dancer, has a wider bandwidth than that available through the existing control channel. Moreover, the actuator that in turn generates the dancer force can be designed to have the necessary wide bandwidth, and operates over a small range of positions. Existing linear actuators are well-suited for this purpose, and an experimental apparatus to verify the simulated behaviour is currently under construction. Finally, we remark that this work is easily generalized to include state-dependent inertias, as in a robotic manipulator.

References

1. M. Spong, "Modeling and control of elastic joint robots," *Transactions of the ASME: Journal of Dynamic Systems, Measurement and Control*, vol. 109, pp. 310–319, December 1987.

2. H. Khalil, *Nonlinear Systems*. Upper Saddle River, NJ: Prentice Hall, 2 ed., 1996.

3. J. Chow and P. Kokotovic, "Two-time-scale feedback design of a class of nonlinear systems," *IEEE Transactions on Automatic Control*, vol. AC-23, pp. 438–443, June 1978.

4. N. Ebler, R. Arnason, G. Michaelis, and N. D'Sa, "Tension control: Dancer rolls or load cells," *IEEE Transactions on Industry Applications*, vol. 29, pp. 727–739, July/August 1993.

5. R. Biekmann, N. Perkins, and A. Ulsoy, "Design and analysis of automotive serpentine belt drive systems for steady state performance," *Transactions of the ASME: Journal of Mechanical Design*, vol. 119, pp. 162–168, June 1997.

6. R. Gorbet and S. Bortoff, "Active compensation of flexibility in belt-driven drive trains," in *Proceedings of the 36th Allerton Conference on Communication, Control, and Computing*, University of Illinois, Urbana, IL, USA, September 1998.

Fault Diagnosis in Finite-State Automata and Timed Discrete-Event Systems *

Shahin Hashtrudi Zad, Raymond H. Kwong, and W. Murray Wonham

Department of Electrical and Computer Engineering, University of Toronto,
Toronto, Ontario, Canada M5S 3G4
Email: hashtrud, kwong, wonham@control.toronto.edu

Abstract. In this paper, we propose a framework for fault diagnosis in discrete-event systems. In this approach, the system and the diagnoser (the fault detection system) do not have to be initialized at the same time. Furthermore, no information about the state or even the condition (failure status) of the system before the initiation of diagnosis is required. First, a state-based approach for on-line passive fault diagnosis in finite-state automata is presented. The design of the fault detection system, in the worst case, has exponential time complexity. A model reduction scheme with polynomial time complexity is introduced to reduce the computational complexity of the design. Next we consider the use of timing information to improve the accuracy of diagnosis. Instead of directly extending our framework to timed discrete-event systems, we take an alternative apporach which leads to significant reduction in on-line computing power requirements and, in many cases, in the size of the diagnoser at the expense of more off-line design calculations. We also discuss the issue of diagnosability of failures in our framework.

1 Introduction

Fault detection systems are of paramount importance in aerospace, manufacturing and process industries. This is due to the crucial role they play in protecting life and property, and in increasing operational time and productivity. Solving diagnostic problems for complex systems is a complicated task requiring a reliable, systematic approach. As a result, fault diagnosis has been the subject of extensive research (see, e.g., [15],[11]).

In this paper, we examine fault diagnosis in systems that, for diagnostic purposes, can be modelled as discrete-event systems (DES). For this problem, different methodologies have been proposed in the literature. By far, fault-tree analysis [9] seems to be the most popular approach. Fault trees can be synthesized automatically. They provide a pictorial display of the system which can be easily read and understood. However, they have certain limitations: it is difficult to incorporate information about ordering and timing of events in a fault tree [11]; moreover, problems arise in the analysis of systems that go through more than one phase of operation [10]; also, there seems to be no way of treating common-cause failures resulting from fault propagation (domino effects) using fault trees [10].

---

* This work was supported in part by AlliedSignal Aerospace Canada.

Other approaches to fault detection proposed by researchers include methods based on artificial intelligence [5] and, templates (see [14] and references therein), event signatures [2] and Petri nets [19] for manufacturing processes.

Finite-state automata have also been used in the study of diagnostic problems. In [12], a state-based approach is utilized for the study of off-line diagnosis. Here, the state of the system is assumed fixed during diagnostic tests, and output measurements are used for failure detection and isolation. The concept of testability is also introduced and studied. An active on-line diagnosis problem has been studied in [12]. In this case, input test sequences for fault diagnosis are computed. Furthermore, the notion of on-line diagnosability is introduced. In [16], [17] passive on-line diagnosis is studied using an event-based framework: based on observed events, inference is made about the occurrence of (unobservable) failure events. Two different notions of diagnosability are defined and examined. This framework has been extended to utilize information about the timing of events [3]. Integrated approaches to fault detection and supervisory control have also been studied in an event-based framework in [4] and [18].

In this paper, we first present a state-based approach for passive on-line diagnosis of finite-state automata. Engineers have found that the use of finite-state automata and the corresponding state transition diagrams make the design and maintenance of complex control and diagnostic systems easier (see, e.g., Ch. 14 of [11]). It is also straightforward to capture the ordering of events using finite-state automata. There are similarities between our work and the event-based approach of [16]. However, our framework is simpler and the diagnoser is easier to compute. Moreover, we do not assume that the plant and the diagnoser are initialized at the same time. Knowledge of the state and even the condition of the plant (normal/faulty) at the time the diagnoser is initialized is not required. It can be shown [6] that any problem in an event-based framework can be recast as a problem in the state-based framework presented here. We also introduce a model reduction scheme with polynomial time complexity to reduce the computational complexity of designing the diagnoser, which has exponential time complexity in the worst case. To our knowledge, the use of model reduction in the design of diagnoser has not been studied previously in DES literature.

Next we discuss the extension of our framework for incorporating timing information. In this case, it is assumed that the system is modelled as a timed discrete-event system (TDES). First introduced in [1] in the study of supervisory control, timed discrete-event systems can be used for capturing timing information in a useful range of problems in control engineering in a reasonably simple fashion. For the purpose of fault diagnosis in TDES, it is possible to adopt our diagnosis theory for finite-state automata simply by treating the clock tick as an extra output signal. In this work, however, we have taken an alternative approach in which the process of updating the estimate of the system's condition is performed only when a new output symbol is generated. The update process uses the generated output symbols and the number of clock ticks between them. No update at clock ticks is required in this method. This results in significant reduction in on-line computing power requirements and, in many cases, in the size of the diagnoser, at the expense of extra off-line design calculations.

We note that testing of finite-state machines [8] is related to fault diagnosis. However, the framework used for that purpose is different: the finite-state machines are usually assumed to be deterministic with a fixed condition (failure status); also it is assumed that transitions can always be observed even if they do not result in a change in output. These assumptions often do not hold in fault diagnosis of control systems.

In this paper, we shall present an overview of our framework and illustrate it using examples. Some of the technical details can be found in [6]. A complete account is given in [7]. An outline of the paper is as follows. In sect. 2 we study failure modelling, introduce our framework for the design of fault diagnosis system and discuss failure diagnosability in discrete-event systems. We propose an extension of our approach for diagnosis in timed discrete-event systems in sect. 3. Section 4 presents the conclusion.

2   Fault Diagnosis in Finite-State Automata

In this section, we study fault diagnosis in finite-state automata.

2.1   Plant Model

We assume that the plant under control, i.e., plant along with low-level continuous controllers and DES supervisors, can be modelled as a finite-state Moore automaton $G = (X, \Sigma, \delta, x_0, Y, \lambda)$, where X, Σ, Y are the finite state, event and output set; x_0 is the initial state, $\delta : X \times \Sigma \to 2^X$ the transition function and $\lambda : X \to Y$ the output map (2^X denotes the power set of X).

The model describes the behaviour of the system in both normal (system functioning properly) and faulty situations. Suppose there are p *failure modes* $F_1, \cdots, F_p$. Each failure mode corresponds to some kind of failure in an instrument (valve, sensor, etc.). The event set Σ includes *failure events*. In this paper, for brevity, we assume at most one failure mode may occur at a time. Simultaneous occurrence of two (or more) failure modes is discussed in [7].

Let $\mathcal{K} := \{N, F_1, \cdots, F_p\}$ denote the *condition* set of the system. It is assumed that the state set X can be partitioned according to the condition of the system: $X = X_N \dot\cup X_{F_1} \dot\cup \cdots \dot\cup X_{F_p}$ ($\dot\cup$ denotes disjoint union). Define $\kappa : X \to \mathcal{K}$ such that for every $x \in X$, $\kappa(x)$ is the condition of the system at the state x: $\kappa(x) = N$ if $x \in X_N$, and $\kappa(x) = F_i$ if $x \in X_{F_i}$ ($i \in \{1, \cdots, p\}$). Also (abusing notation) extend the definition of κ to the subsets of X: $\kappa(z) = \cup\{\kappa(x) \mid x \in z\}$, for any $z \subseteq X$.

In failure detection and isolation, given the output sequence $(y_1, y_2, y_3, \cdots)$, we want to find the condition of the system. In general, some of the events in Σ are observable; it is assumed that information about the occurrence of these events has been transferred and included in the output map. Also note that only changes in the output are assumed to be observable. This means that a transition of the system from one state to another state having the same output will not be noticed, i.e., the transition will be unobservable. So in the output sequence: $y_i \neq y_{i+1}$ for $i \geq 1$.

Example 1 - Heating System

A heating system uses a heater, a temperature sensor and an ON/OFF controller to regulate the temperature of a room about a set-point. The DES model is shown in Fig. 1. In this figure, each dashed arc represents a heater-failure event. "Load" models the effect of disturbance such as the temperature of the adjoining room and the ambient temperature, and is supposed to have two states "normal (n)" and "above normal (a)". It is also assumed that even when the load is above normal, the heater can keep the temperature close to the set-point. In the heating system: $X = \{1, 2, \cdots, 24\}$, $Y = \{ld, le, bd, be, ad, ae\}$,

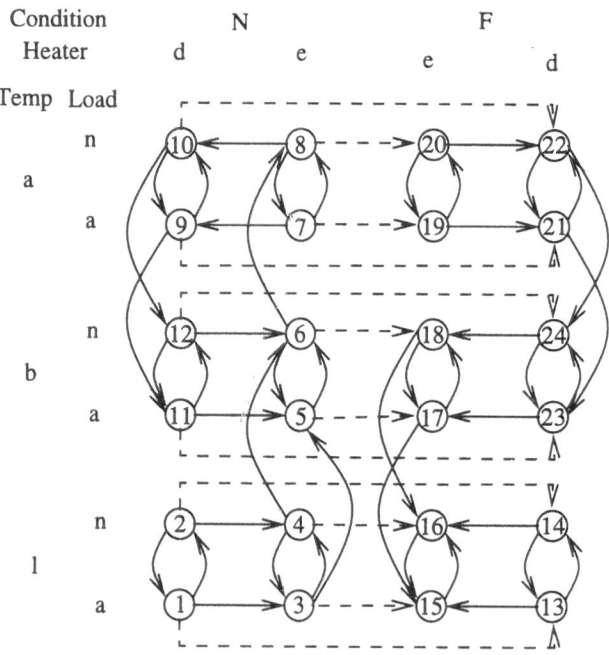

Fig. 1. Example 1. DES model of the heating system.

$\mathcal{K} = \{N, F\}$, $\kappa(i) = N$ for $1 \leq i \leq 12$, and $\kappa(i) = F$ for $13 \leq i \leq 24$. The output symbols are explained below:

ld temperature low, heater OFF (disabled)
le temperature low, heater ON (enabled)
bd temp. below set-point, heater OFF (disabled)
be temp. below set-point, heater ON (enabled)
ad temp. above set-point, heater OFF (disabled)
ae temp. above set-point, heater ON (enabled).

Note that the output symbol may contain information about the commands issued by the DES supervisor (in this case, the ON/OFF controller), in addition to sensor readings.

2.2ˋ Diagnoser Construction and Diagnosability

Before studying diagnoser design, we introduce the concept of output-adjacency.

Definition 1. For any two states $x, x' \in X$, we say x' is *output-adjacent* to x and write $x \Rightarrow x'$ if $\lambda(x) \neq \lambda(x')$ and there exist $l \geq 2$, $x_2, \cdots, x_{l-1}$, $\sigma_1, \cdots, \sigma_{l-1}$ such that $x_{i+1} \in \delta(x_i, \sigma_i)$ and $\lambda(x_i) = \lambda(x)$, for all $1 \leq i \leq l-1$, with $x_1 = x$ and $x_l = x'$.

A *diagnoser* is a system that detects and isolates failures. In our framework, it is a finite-state Moore machine that takes the output sequence of the system $(y_1, y_2, \cdots, y_k)$ as input and generates at its output an estimate of the condition of the system at the time that y_k was generated. Specifically, based on the output sequence up to y_k, a set $z_k \in 2^X - \{\emptyset\}$ is calculated to which x can belong at the time that y_k was generated. $\kappa(z_k)$ will be the estimate of the system's condition (Fig. 2). Upon observing y_{k+1}, z_k will be updated to z_{k+1}. Therefore z_{k+1} (for $k \geq 1$), as depicted in Fig. 3, will be the set of states,

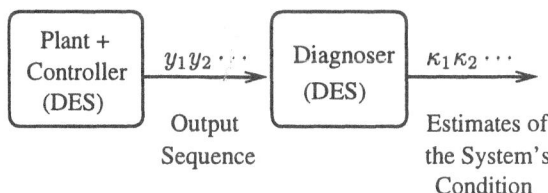

Fig. 2. System and diagnoser.

having output y_{k+1} that are reachable from the states in z_k using paths along which the output is y_k.

We denote the initial state estimate by z_0. It contains the information available about the state of the system at the time that the diagnoser is initialized, *before* the reading of sensors begins. Usually, $z_0 = X$, because the diagnoser may be initialized at any time while the system is in operation and in this situation the state of the system is not known exactly. If the system

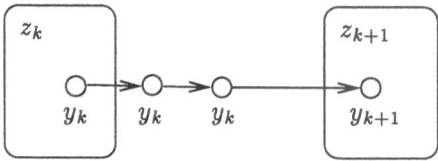

Fig. 3. $z_{k+1} = \zeta(z_k, y_{k+1})$

and the diagnoser are initialized at the same time, then $z_0 = \{x_0\}$. If the system is only known to be normal at the time that the diagnoser is initialized, then $z_0 = X_N$. $z_1 = z_0 \cap \lambda^{-1}(\{y_1\})$ is the first estimate of the system's state immediately after observing the first output symbol y_1. For $k \geq 1$, $z_{k+1} = \{x \mid \lambda(x) = y_{k+1}$ & $(\exists\, x' \in z_k : \; x' \Rightarrow x)\}$. The transition map of the diagnoser is denoted by ζ.

Now we introduce the *reachability transition system* (RTS). The RTS will be useful in the computation of diagnoser, in on-line implementation of diagnostic computations and in model reduction. Suppose $z^1, z^2 \in Z$ and $\zeta(z^1, y) = z^2$ for some $y \in Y$. Given z^1 and in order to compute z^2, we have to find for every $x_1 \in z^1$, all x_2 such that $\lambda(x_2) = y$ and $x_1 \Rightarrow x_2$. Since every $x \in X$ typically belongs to several states of the diagnoser, it is computationally economical to compute the set of output-adjacent states of every $x \in X$. This can be done in $\mathcal{O}(|X|^2 + |X||T|)$ time because a breadth-first search reachability analysis for each $x \in X$ can be done in $\mathcal{O}(|X| + |T|)$ time. Here $|X|$ and $|T|$ are the cardinalities of X and T (the set of transitions of G). Having the sets of output-adjacent states, we can construct the reachability transition system $\tilde{G} = (X, R, Y, \lambda)$, which has X, Y and λ as the state set, output set and output map. $R \subseteq X \times X$ is a binary relation and $(x_1, x_2) \in R$ if and only if $x_1 \Rightarrow x_2$. As mentioned above, $\tilde{G}$ can be computed in $\mathcal{O}(|X|^2 + |X||T|)$ time. With $\tilde{G}$ available, one can compute the diagnoser.

Example 1 - Heating System (Cont'd)
The RTS (in the form of a table) and the diagnoser for the heating system are given in Table 1 and Fig. 4 (assuming $z_0 = X$). In this example, heater failure can be detected using output observations unless it occurs while the temperature is low. For example, suppose that the diagnoser is initialized when the system is at the state 10. The output at this state is 'ad'. Therefore $z_1 = \{9, 10, 21, 22\}$ and $\kappa(z_1) = \{N, F\}$. Now assume that the state x moves to 12, and then to 6; at this point a heater failure occurs following which the state moves to 18 and finally to 16. While the system evolves along this path, the output sequence 'bd, be, le' will be generated which will take the diagnoser to the state $z_4 = \{15, 16\}$. $\kappa(z_4) = \{F\}$; thus the diagnoser eventually detects the failure.

To see why heater failure cannot be detected while the temperature is low, suppose the system is at the state 3 when the diagnoser is started. At this state the output is 'le'; thus $z_1 = \{3, 4, 15, 16\}$ and $\kappa(z_1) = \{N, F\}$. A heater failure at this point takes the state x to 15. No new output symbol will be generated following the failure or afterwards. Therefore, $\{N, F\}$ will remain as the estimate of the system's condition. As a result, the failure will not be detected with certainty.

Table 1. Example 1. Reachability transition system.

State	Output-adjacent states (output)	State	Output-adjacent states (output)
1	3,4,15,16 (le)	13	15,16 (le)
2	3,4,15,16 (le)	14	15,16 (le)
3	5,6 (be)	15	-
4	5,6 (be)	16	-
5	8 (ae) / 15,16 (le)	17	15,16 (le)
6	8 (ae) / 15,16 (le)	18	15,16 (le)
7	9,10,21,22 (ad)	19	21,22 (ad)
8	9,10,21,22 (ad)	20	21,22 (ad)
9	11,12,23,24 (bd)	21	23,24 (bd)
10	11,12,23,24 (bd)	22	23,24 (bd)
11	5,6,17,18 (be)	23	17,18 (be)
12	5,6,17,18 (be)	24	17,18 (be)

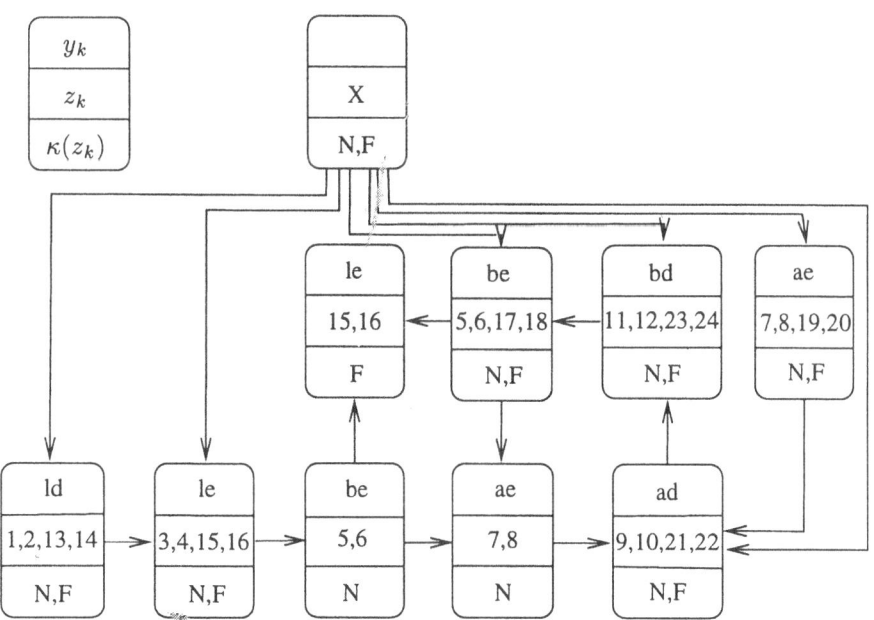

Fig. 4. Example 1. Diagnoser.

In the example, after the occurrence of failure, the heating system does not return to the normal condition, i.e., the failure is *permanent* or *hard*. In control, we are usually interested in permanent failures. In the remainder of this subsection, we assume the failure modes are all permanent; non-permanent failures are discussed in [7].

Consider a permanent failure mode F_i. A state of the diagnoser z is called F_i-*certain* if $\kappa(z) = \{F_i\}$, i.e., $z \subseteq X_{F_i}$. If the diagnoser enters an F_i-certain state, it means that F_i has been detected and isolated. If $F_i \in \kappa(z)$ but $\kappa(z) \neq \{F_i\}$, then z is called F_i-*uncertain*. If for some $k_0 \geq 0$ and a failure F_i, $\kappa(z_{k_0}) = \{F_i\}$, then $\kappa(z_k) = \{F_i\}$ for all $k \geq k_0$ (because $X_N \cup (\cup_{j \neq i} X_{F_j})$ is not reachable from X_{F_i}).

Definition 2. A permanent failure mode F_i is *diagnosable* if there exists an integer $N \geq 0$ such that following both the occurrence of the failure and initialization of the diagnoser, F_i can be detected and isolated (i.e., the diagnoser reaches an F_i-certain state) after the occurrence of at most N events in the system.

In the above definition, no assumption is made about the system's condition (normal/faulty) at the time that the diagnoser is initialized, i.e., the diagnoser might have been initialized either before or after the occurrence of the failure. According to Def. 2, the heater failure in Example 1 is not diagnosable.

Necessary and sufficient conditions for diagnosability of permanent failures are obtained in [7]. Intuitively, a failure would not be diagnosable if after it occurs, no new output symbols are generated. Also a failure mode F_i would be undiagnosable if in this mode, the system can generate a periodic output sequence that throws the diagnoser into a cycle of F_i-uncertain states.

Since the number of states of the diagnoser is in the worst case exponential in $|X|$, the whole diagnoser may occupy a large space in computer memory. In this case, it might be better to store the reachability transition system in memory and use it to perform the diagnostic computations on-line: having z_k and the observation y_{k+1}, use $\tilde{G}$ to compute z_{k+1} and update the estimate of the system's condition. $\tilde{G}$ contains $|X|$ states and at most $|X|(|X| - 1)$ transitions.

There are similarities between our framework and that of [16], especially in the use of an observer for diagnosis. However, while we try to determine the *condition* of the system for fault detection, in [16] the authors attempt to detect the unobservable failure *events*. This leads to some differences. (i) In our approach, the state and even the condition of the system do not have to be known at the time that the diagnoser is started. The diagnoser can be initialized at any time while the system is in operation, not necessarily when the system is started. If say a failure occurs before the diagnoser is initialized, then our diagnoser can eventually detect the faulty condition and isolate the failure (assuming the failure is diagnosable). However, an event-based diagnoser cannot detect the failure because the failure event has already happened when the diagnoser is initialized. (ii) Our approach is simpler and in this framework, computation of the diagnoser is less complex. This is because at each step, after observing a new output symbol, we only have to update our estimate of the system's state z_k. In the event-based approach of [16], after the occurrence of an observable event, in addition to updating the state estimate, all of the paths that the state of the system might have evolved along since the occurrence of the previous observable event, have to be checked for the occurrence of failure events.

We note that a diagnosis problem in an event-based framework can always be transformed to an equivalent problem in the state-based framework presented in this paper [6].

2.3   Model Reduction

The computational complexity of designing the diagnoser in the worst case is exponential in the number of system states. To mitigate this, in this section we introduce a model reduction scheme with polynomial time complexity. The equivalence relation used for model reduction is based on the solution of the *relational coarsest partition* (RCP) problem for the reachability transition system. We will show that the diagnoser built using the reduced RTS will be *equivalent* to the original diagnoser in the following sense.

Definition 3. Two diagnosers for a system are *equivalent* if for any given output sequence, they produce identical sequences of estimates for the system's condition.

Consider the RTS $\tilde{G} = (X, R, Y, \lambda)$. For every $x_1, x_2 \in X$, let $x_2 \in R(x_1)$ iff $(x_1, x_2) \in R$, i.e., $x_1 \Rightarrow x_2$. Let $\pi = \{B_1, \cdots, B_{|\pi|}\}$ be a partition of X, with B_i denoting the blocks of π. Then π is *compatible* with R iff whenever x and x' are in the same block B_i, then for any block B_j, $R(x) \cap B_j \neq \emptyset$ iff $R(x') \cap B_j \neq \emptyset$. Let Π be the set of partitions compatible with R.

Suppose π is compatible with R and $\pi \leq \ker\lambda \wedge \ker\kappa$, where ker refers to the equivalence kernel of the corresponding map and $\wedge$ denotes the meet operation in the lattice of equivalence relations. If two states x and x' are in the same block of $\ker\lambda \wedge \ker\kappa$, then the system has the same output and condition at these states. Moreover, the set of output sequences generated by the system, starting at either of these states, will be identical. Also, starting at any of these states and for any feasible output sequence generated, the sequence of condition estimates will be the same. This shows that the three statements $x \in z$, $x' \in z$ and $x, x' \in z$ contain the same information about the present and future estimates of the system's condition. Hence, for the purpose of estimating the system's condition, x and x' are equivalent.

Let $P : X \to X/\pi$ be the canonical projection. For every $x \in X$, $[x] := Px$ denotes the block x belongs to. Also for simplicity, instead of $x \in P^{-1}\bar{x}_1$, we write $x \in \bar{x}_1$. We define the *reduced RTS* $\overline{G} = (\overline{X}, \overline{R}, Y, \overline{\lambda})$ (corresponding to π) according to: (i) $\overline{X} = X/\pi$; (ii) for all $\bar{x}_1, \bar{x}_2 \in \overline{X}$:

$$(\bar{x}_1, \bar{x}_2) \in \overline{R} \Leftrightarrow (\forall\, x \in \bar{x}_1\ \exists\, x' \in \bar{x}_2 : \ (x, x') \in R);$$

and (iii) for all $\bar{x} \in \overline{X}$: $\overline{\lambda}(\bar{x}) = \lambda(x)$ for any $x \in \bar{x}$. Similarly we define $\bar{\kappa} : \overline{X} \to \mathcal{K}$ according to $\bar{\kappa}(\bar{x}) = \kappa(x)$ for any $x \in \bar{x}$. Since $\pi \leq \ker\lambda \wedge \ker\kappa$, $\overline{\lambda}$ and $\overline{\kappa}$ are well-defined. We define the canonical projection of a subset $z \subseteq X$ to be $P(z) := \bigcup\{[x] | x \in z\}$.

We refer to the diagnoser designed based on the reduced RTS $\overline{G}$ with $\bar{z}_0 := P z_0$ as its initial state estimate, as the *high-level diagnoser* (correspond-ing to π) and, call it $\overline{D}$.

Theorem [7] *The original diagnoser and the high-level diagnoser are equiva-lent.*

Let $\bar{z}_k$ and $\bar{\zeta}$ denote the state and transition function of $\overline{D}$.

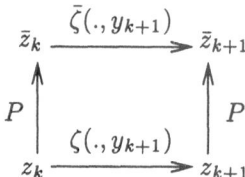

Fig. 5. $P z_k = \bar{z}_k$

Remark 1. It can be shown [7] that $P z_k = \bar{z}_k$ (Fig. 5). Therefore since P is onto, the number of states of $\overline{D}$ is less than or equal to that of D.

Remark 2. Obviously the coarser the partition π is, the fewer states will the reduced RTS have. The set $\{\pi \in \Pi, \; \pi \leq \ker\lambda \wedge \ker\kappa\}$ has a unique supre-mal element which is the coarsest partition compatible with R and finer than $\ker\lambda \wedge \ker\kappa$. The problem of computing this supremal element is called the *relational coarsest partition* (RCP) problem. The efficient algorithm of [13] solves the RCP problem in $\mathcal{O}(|R| \log |X|)$ time ($|R|$ is the cardinality of the set R). This gives a $\mathcal{O}(|X|^2 \log |X|)$ time complexity for model reduction since $|R| \leq |X|(|X| - 1)$.

Remark 3. Sometimes the output map λ provides redundant information, i.e., fault diagnosis can be accomplished using a coarser output map. The use of a coarser output map, in general, results in further aggregation in model reduction, hence fewer states in the reduced RTS.

Remark 4. The high-level diagnoser produces an accurate estimate of the sys-tem's condition, with an accuracy proportionate to the amount of information available in the reduced RTS, iff the relation $P z_k = \bar{z}_k$ holds. It can be shown [7] that $P z_k = \bar{z}_k$ holds for a partition $\pi \leq \ker\lambda \wedge \ker\kappa$ if π is compatible with R. If π is not compatible with R, then the high-level state estimates will become either conservative, i.e., $P z_k \subseteq \bar{z}_k$ or risk-accepting, i.e., $\bar{z}_k \subseteq P z_k$ [7]. Thus the use of compatible partitions is necessary for guaranteeing $P z_k = \bar{z}_k$ and in this case, the coarsest partition compatible with R and finer than $\ker\lambda \wedge \ker\kappa$ will be the optimal partition for model reduction, i.e., it will give the reduced RTS with minimum number of states.

Example 1 - Heating System (Cont'd)

Using Table 1, the reader can verify that the partition $\ker\lambda \wedge \ker\kappa = \{\{1,2\},$ $\{3,4\}, \{5,6\}, \cdots, \{23, 24\}\}$ is compatible with the transition relation R of the RTS. Hence it is the solution to the RCP problem for this example. The reduced RTS is given in Table 2 where each pair of states $2i - 1$ and $2i$ ($1 \leq i \leq 12$) in the original RTS is replaced with i' in the reduced RTS. Therefore the number of states of the RTS has been reduced to half. The high-level diagnoser is identical to the original one (After replacing each pair of $2i - 1$ and $2i$ with i').

Table 2. Example 1. Reduced reachability transition system.

State	Output-adjacent states (output)	State	Output-adjacent states (output)
1'	2',8' (le)	7'	8' (le)
2'	3' (be)	8'	-
3'	4' (ae) / 8' (le)	9'	8' (le)
4'	5',11' (ad)	10'	11' (ad)
5'	6',12' (bd)	11'	12' (bd)
6'	3',9' (be)	12'	9' (be)

The fact that the pair of states $2i$ and $2i - 1$ are equivalent and can be replaced by the single state i', shows that incorporating the model of the load has not added any useful information to the heating system's model for the purpose of fault diagnosis and therefore, the load model can be removed. This is an interesting point and, is useful for decentralized fault diagnosis. Think of the load as subsystem 1 and of the rest of the heating system as subsystem 2. Our model reduction scheme has shown that in the absence of sensor readings from subsystem 1 (which is typical in a decentralized fault detection scheme), the model of subsystem 2 (the local model) has the same amount of information for fault diagnosis as the combined model of the subsystems (the centralized model). Thus the design of diagnoser for subsystem 2 can be done based on the local model only.

3   Fault Diagnosis in Timed Discrete-Event Systems

In sect. 2, changes in the output sequence resulting from failures were used for fault diagnosis in finite-state Moore automata. In many cases, however, a failure does not always change the output sequence; it only changes the *timing* of the output sequence. In other cases, as a result of failure, the system stops generating new output symbols (e.g., when the temperature is low in Example 1). In these situations, timing information may help us to detect and isolate failures. The use of timing information increases the accuracy of the fault detection system at the expense of additional computational complexity.

3.1 Plant Model

In this section, we assume that the plant under control can be modelled as a timed discrete-event system [1]. A TDES is a finite-state automaton . One of the events (in the event set of the TDES) is the tick of a global clock. The tick event is assumed to be observable. The TDES model can be used to describe the sequence of events occurring in the system with respect to the ticks of the global clock. The TDES model used in this paper is similar to that in [1] except that our model is nondeterministic and, it includes an output and a condition map defined on its state set. For further details, the reader is referred to [7]. Here we explain the model using the following example.

Example 2 - Neutralization Process
In a (simplified) neutralization process (Fig. 6) all valves are closed and the tank is almost empty initially. The process starts by opening valve V1 and

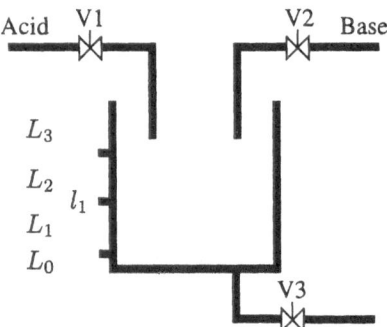

Fig. 6. Example 2. A neutralization process.

filling the reaction tank up to level l_1 with the chemical to be treated (here acid). Next, V1 is closed and the neutralizer (base) is added by opening valve V2. When the alkalinity (pH) of the solution reaches the normal range, V2 is closed and the tank contents are drained through valve V3. This completes a cycle of the process. Following this, another cycle will be started.

The events and the corresponding time bounds (to be explained later) are given in Table 3. Two failure modes are considered here: valve V1 stuck open (F_1) and valve V1 stuck closed (F_2). For simiplcity, it is assumed V1 gets stuck open (resp. closed) only when it is open (resp. closed). Sensor measurements are 'pH' and 'Level', with pH∈ $\{a(\text{acid}), n(\text{neutral}), b(\text{base})\}$ and Level∈ $\{L_0, L_1, L_2, L_3\}$.

Table 3. Example 2. Events and their time bounds.

Event		Time bounds
oi	valve i open	[1,1]
ci	valve i closed	[1,1]
L_{ij}	level change from L_i to L_j	$[l', u']$ for L_{23} $[l, u]$ otherwise
a2n	pH change from a (acid) to n (neutral)	$[l,u]$
n2a	pH change from n (neutral) to a (acid)	[1,1]
W2	wait for 2 clock ticks	2
F_1	valve V1 stuck open	$[0, \infty)$
F_2	valve V1 stuck closed	$[0, \infty)$

The control sequence consists of 8 steps:

C1: Order V1 open;
   WAIT UNTIL pH=a OR Level=L_1;
   IF pH=a GO TO C2;
   ELSE GO TO C3
C2: WHEN Level=L_1 GO TO C4
C3: WHEN pH=a GO TO C4
C4: WHEN Level=L_2 GO TO C5
C5: Order V1 closed;
   Order V2 open;
   WHEN pH=n GO TO C6
C6: Order V2 closed;
   Order V3 open;
   WHEN Level=L_1 GO TO C7
C7: WHEN Level=L_0 GO TO C8
C8: Order V3 closed;
   WAIT for 2 clock ticks;
   GO TO C1

Note that by assumption, the controller does not have a way of knowing whether a valve is open or closed. It is also assumed that in addition to Level and pH, the current step in the control sequence is known to the diagnoser. Therefore $Y \subseteq \{C1, \cdots, C8\} \times \{L_0, L_1, L_2, L_3\} \times \{a, n, b\}$. The output and the condition map are given in Table 4.

The TDES describing the neutralization process is depicted in Fig. 7 (assuming $l = 2$, $u = 3$, $l' = 1$ and $u' = 2$ for the time bounds of the L_{ij} as given in Table 3). It consists of three subautomata each describing the behaviour of the system in the N or F_1 or F_2 condition. Note that, to avoid cluttering the graph, we have depicted the transitions from the normal mode N to the faulty conditions F_1 and F_2 (i.e., failure events) only on the subautomata corresponding to the faulty conditions. In Fig. 7, τ is the tick event. Initially, the system is at $x_0 = 1$ where the controller enables the event o1 (open V1). The

Table 4. Example 2. Output and condition map.

State	Output, Condition	State	Output, Condition
1	$(C1,L_0,n)$, N	$2.1'$	$(C1,L_0, n)$, F_1
1.1	$(C1,L_0,n)$, N	$2.2'$	$(C1,L_0, n)$, F_1
2	$(C1,L_0,n)$, N	$3'$	$(C2,L_0,a)$, F_1
2.1	$(C1,L_0,n)$, N	$4'$	$(C3,L_1,n)$, F_1
3	$(C2,L_0,a)$, N	$5'$	$(C4,L_1,a)$, F_1
4	$(C3,L_1,n)$, N	$5.1'$	$(C4,L_1,a)$, F_1
5	$(C4,L_1,a)$, N	$5.2'$	$(C4,L_1,a)$, F_1
5.1	$(C4,L_1,a)$, N	$5.3'$	$(C4,L_1,a)$, F_1
5.2	$(C4,L_1,a)$, N	$6'$	$(C5,L_2,a)$, F_1
5.3	$(C4,L_1,a)$, N	$6.1'$	$(C5,L_2,a)$, F_1
6	$(C5,L_2,a)$, N	$8'$	$(C5,L_2,a)$, F_1
6.1	$(C5,L_2,a)$, N	$8.1'$	$(C5,L_2,a)$, F_1
7	$(C5,L_2,a)$, N	$8.2'$	$(C5,L_2,a)$, F_1
8	$(C5,L_2,a)$, N	$17'$	$(C5,L_3,a)$, F_1
9	$(C5,L_2,a)$, N	$7''$	$(C5,L_2,a)$, F_2
9.1	$(C5,L_2,a)$, N	$9''$	$(C5,L_2,a)$, F_2
9.2	$(C5,L_2,a)$, N	$9.1''$	$(C5,L_2,a)$, F_2
9.3	$(C5,L_2,a)$, N	$9.2''$	$(C5,L_2,a)$, F_2
10	$(C6,L_2,n)$, N	$9.3''$	$(C5,L_2,a)$, F_2
10.1	$(C6,L_2,n)$, N	$10''$	$(C6,L_2,n)$, F_2
11	$(C6,L_2,n)$, N	$10.1''$	$(C6,L_2,n)$, F_2
12	$(C6,L_2,n)$, N	$11''$	$(C6,L_2,n)$, F_2
13	$(C6,L_2,n)$, N	$12''$	$(C6,L_2,n)$, F_2
13.1	$(C6,L_2,n)$, N	$13''$	$(C6,L_2,n)$, F_2
13.2	$(C6,L_2,n)$, N	$13.1''$	$(C6,L_2,n)$, F_2
13.3	$(C6,L_2,n)$, N	$13.2''$	$(C6,L_2,n)$, F_2
14	$(C7,L_1,n)$, N	$13.3''$	$(C6,L_2,n)$, F_2
14.1	$(C7,L_1,n)$, N	$14''$	$(C7,L_1,n)$, F_2
14.2	$(C7,L_1,n)$, N	$14.1''$	$(C7,L_1,n)$, F_2
14.3	$(C7,L_1,n)$, N	$14.2''$	$(C7,L_1,n)$, F_2
15	$(C8,L_0,n)$, N	$14.3''$	$(C7,L_1,n)$, F_2
15.1	$(C8,L_0,n)$, N	$15''$	$(C8,L_0,n)$, F_2
16	$(C8,L_0,n)$, N	$15.1''$	$(C8,L_0,n)$, F_2
16.1	$(C8,L_0,n)$, N	$16''$	$(C8,L_0,n)$, F_2
		$16.1''$	$(C8,L_0,n)$, F_2
		$1''$	$(C1,L_0,n)$, F_2
		$1.1''$	$(C1,L_0,n)$, F_2

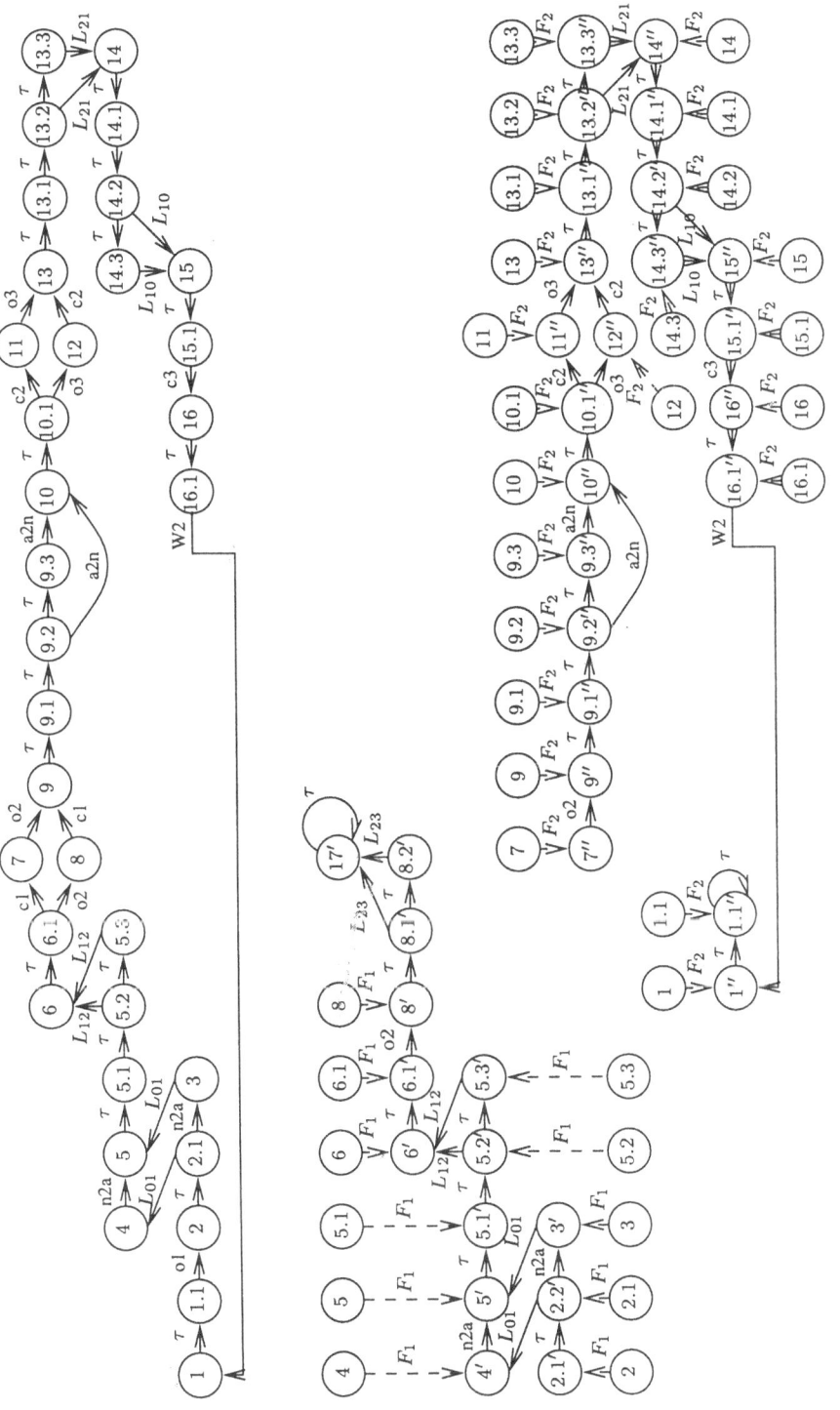

Fig. 7. Example 2. TDES model.

lower and upper bounds of o1 are both 1 which means that following its enablement, o1 cannot occur before the first clock tick but if it remains enabled, it will certainly happen before the second clock tick. In general, a time bound $[a, b]$ for an event σ means that after σ is enabled, it cannot occur before a ticks of the clock but if it remains enabled, it will occur before the $(b + 1)$-th tick (following its enablement). Reverting to the TDES, the event o1 occurs after the first tick but before the second. Following this, the level reaches the L_1 range and the solution becomes acid within the time bounds specified in Table 3. Then level reaches L_2 in $l(= 2)$ to $u(= 3)$ clock ticks and, so on. Note that at state 15 when the controller orders V3 closed, it waits two ticks of the clock before starting a new cycle to make sure V3 is closed at the start of the new cycle. While V1 is open, it may fail stuck-open which ultimately results in the occurrence of L_{23}. Also it may fail stuck-closed while it is closed. In this case, at the beginning of the new cycle when the controller orders V1 open, neither L_{01} or n2a will happen.

A TDES is usually obtained from an activity transition graph (ATG) [1]. ATG models are more convenient for describing discrete-event processes. In this paper, we do not discuss the ATG of the neutralization process for the sake of brevity. The details are given in [7].

For diagnoser design, it is useful to project out the clock ticks [1] from the TDES. We refer to the resulting system which describes the order and duration (in ticks) of occurrence of the non-tick events in the plant as the *timed finite-state Moore automaton* (corresponding to the TDES model). Let Q_τ, Σ_τ, δ_τ, q_0, X, Σ, δ and x_0 denote the state set, event set, transition function and initial state of the TDES and the timed FSMA, respectively. Then

$$X = \{x \in Q_\tau | \exists \sigma \in \Sigma_\tau - \{\tau\}, x' \in Q : \quad x \in \delta_\tau(x', \sigma)\} \cup \{q_0\},$$
$$\Sigma = \Sigma_\tau - \{\tau\},$$
$$x_0 = q_0.$$

Notation For every $x, x' \in X$ and $\sigma \in \Sigma$, $x \overset{\sigma}{\Rightarrow} x'$ iff there exist $l \geq 2$, $q_1, \cdots, q_l$, with $q_1 = x$ and $q_l = x'$, such that $q_{i+1} \in \delta_\tau(q_i, \tau)$ for $1 \leq i \leq l - 2$ and $q_l \in \delta(q_{l-1}, \sigma)$.

The transition function $\delta : X \times \Sigma \to 2^X$ is defined according to:

$$\forall x_1, x_2 \in X, \sigma \in \Sigma : \quad x_2 \in \delta(x_1, \sigma) \quad \Leftrightarrow \quad x_1 \overset{\sigma}{\Rightarrow} x_2.$$

By definition, for any $x \in X$, the output $\lambda(x)$ and condition $\kappa(x)$ in the timed FSMA are the same as those in the TDES. We also define the *transition time*

---

[1] In the process of projecting out the τ transitions, it is assumed that no τ transition results in a change in the output (because if it does, the information about the output change will be lost after projection). If this is not the case, then, before projection, any transition $x_1 \overset{\tau}{\to} x_2$ with $\lambda(x_1) \neq \lambda(x_2)$ should be replaced with the two transitions $x_1 \overset{\tau}{\to} x_1' \overset{\sigma_{12}}{\to} x_2$ with $\lambda(x_1') = \lambda(x_1)$, where x_1' and σ_{12} are new state and event that have to be added to the state and event set of the TDES.

function $\mathcal{T} : X \times \Sigma \times X \to 2^{\mathbf{N}}$ ($\mathbf{N} = \{0, 1, 2, \cdots\}$) as follows:

$$\forall x_1, x_2 \in X, \sigma \in \Sigma : \quad n \in \mathcal{T}(x_1, \sigma, x_2) \Leftrightarrow$$

$$\begin{cases} \exists q_1, \cdots, q_n : q_1 \in \delta_\tau(x_1, \tau) \ \& \\ (q_{i+1} \in \delta_\tau(q_i, \tau) \ \forall 1 \le i \le n-1) \ \& \ x_2 \in \delta_\tau(q_n, \sigma) \ \text{if } n \ne 0 \\ \\ x_2 \in \delta_\tau(x_1, \sigma) \hspace{6.5cm} \text{if } n = 0. \end{cases}$$

Obviously if $x_2 \notin \delta(x_1, \sigma)$, then $\mathcal{T}(x_1, \sigma, x_2) = \emptyset$. $\mathcal{T}(x_1, \sigma, x_2)$ is the set of times (in ticks) that a σ-transition from x_1 to x_2 may take in the timed FSMA.

The timed FSMA of the neutralization process is shown in Figures 8 and 9. According to the timed FSMA, for example, the transition from state 5 to

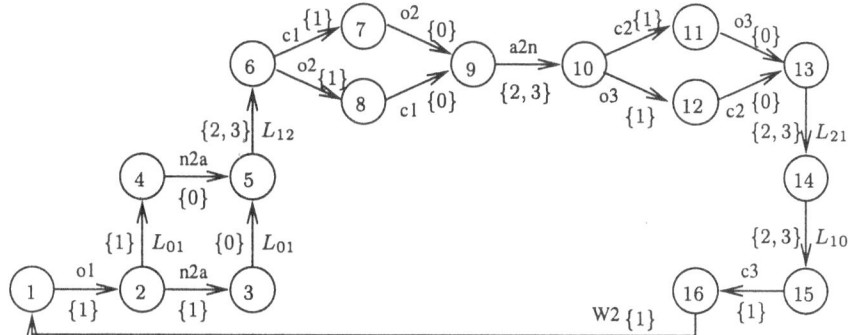

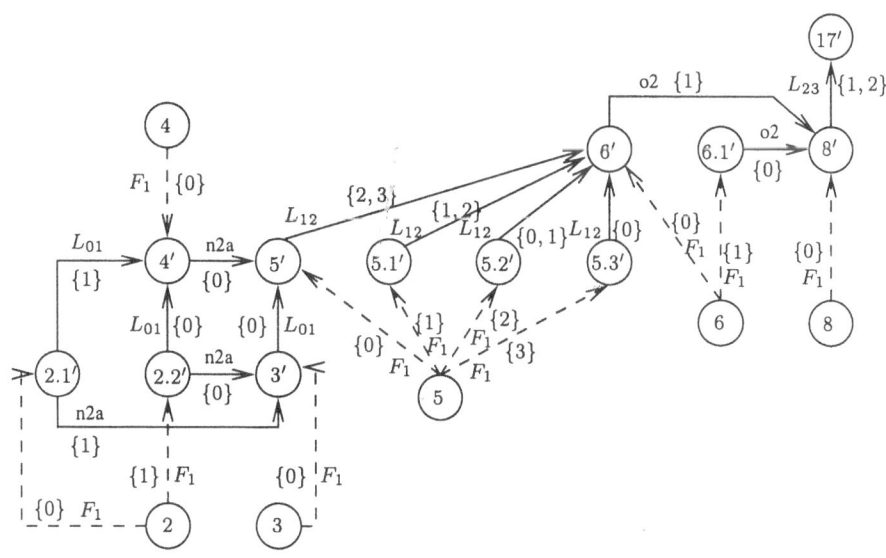

Fig. 8. Example 2. Timed FSMA (Part 1).

6 takes 2 to 3 ticks. This is in accordance with the TDES (Fig. 7).

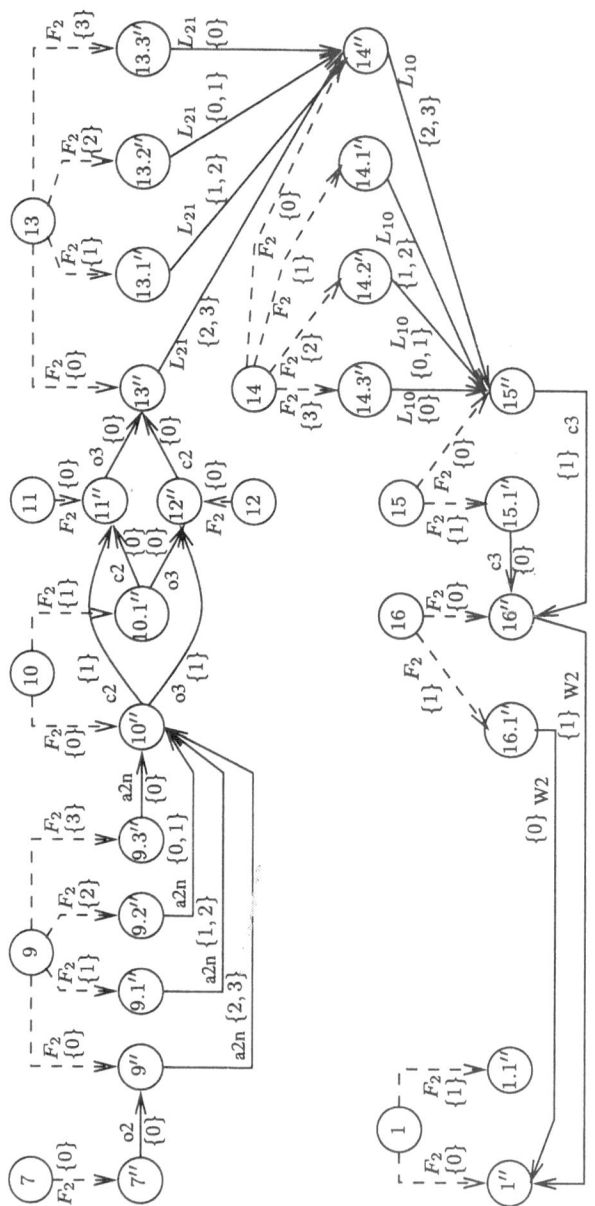

Fig. 9. Example 2. Timed FSMA (Part 2).

3.2   Diagnoser Construction and Diagnosability

For diagnoser design, it is possible to start with the TDES model of the system and treating the clock tick as an extra output signal (i.e., extending the output map to include the ticks), and design a diagnoser using the methodology presented in the previous section. This diagnoser which we will refer to as the *standard diagnoser* provides updates of the system's condition after the generation of any new output symbol in the output set Y and any clock tick. This is a straightforward apporach. However, the number of states of the corresponding RTS and diagnoser will be very large due to the incorporation of timing information.

In this paper, we propose an alternative approach in which the process of updating the estimate of the system's condition is performed only when a new output symbol $y \in Y$ is generated. The update process is based on the generated output symbols and the number of clock ticks occurring between them; no updating at clock ticks is required in this method. This results in significant reduction in on-line computing power requirements and, in many cases, in the size of the diagnoser at the expense of extra off-line design calculations.

Consider the plant under control which in the previous section, was modelled as a timed FSMA. This timed FSMA generates an output sequence $y_1, y_2, \cdots, y_k$. Let t_k denote the number of clock ticks that occurred between the observations y_{k-1} and y_k. The diagnoser proposed in this paper generates a state estimate $z_k \in 2^X - \{\emptyset\}$ for the timed FSMA based on the output sequence $y_1, y_2, \cdots, y_k$ and the timing sequence $t_2, \cdots, t_k$. $\kappa(z_k)$ will be the estimate of the system's condition at the time that y_k was generated. After t_{k+1} ticks and upon observing y_{k+1}, z_k will be updated to z_{k+1}.

As with fault diagnosis in finite-state automata, it is computationally economical to construct a transition system to summarize and store the information about the output-adjacent states of the timed FSMA (The definition of output-adjacent states in timed FSMA is the same as Def. 1). But first define $\tilde{T} : X \times X \to 2^N$ according to

$$
\tilde{T}(x, x') = \begin{cases} \{n \mid n \text{ is the time (in ticks) it can take the timed} \\ \quad \text{FSMA to go from } x \text{ to } x' \text{ using a path along} \\ \quad \text{which the output is } \lambda(x) \text{ (except at } x')\} & \text{if } x \Rightarrow x', \\ \emptyset & \text{otherwise.} \end{cases}
$$

The *timed RTS* corresponding to the timed FSMA of the neutralization process is given in Table. 5, assuming $l = 2$, $u = 3$, $l' = 1$ and $u' = 2$. Note that in this table, for output-adjacent states, the transition time sets $\tilde{T}(.,.)$ are also given. The outputs were provided in Table 4.

The diagnoser for the neutralization process is shown in Fig. 10 (assuming $z_0 = X$). Note that there are 4 parameters in the diagnoser: l, u, l' and u'. The way the diagnoser operates is described in the following. If, for example, the first output symbol is $(C5, L_2, a)$, then $z_1 = \{6, 7, 8, 9, 6', 6.1', 8', 7'', 9'', 9.1'', \cdots, 9.u''\}$. If the next output symbol y_2 is $(C5, L_3, a)$, then $z_2 = \{17'\}$ and

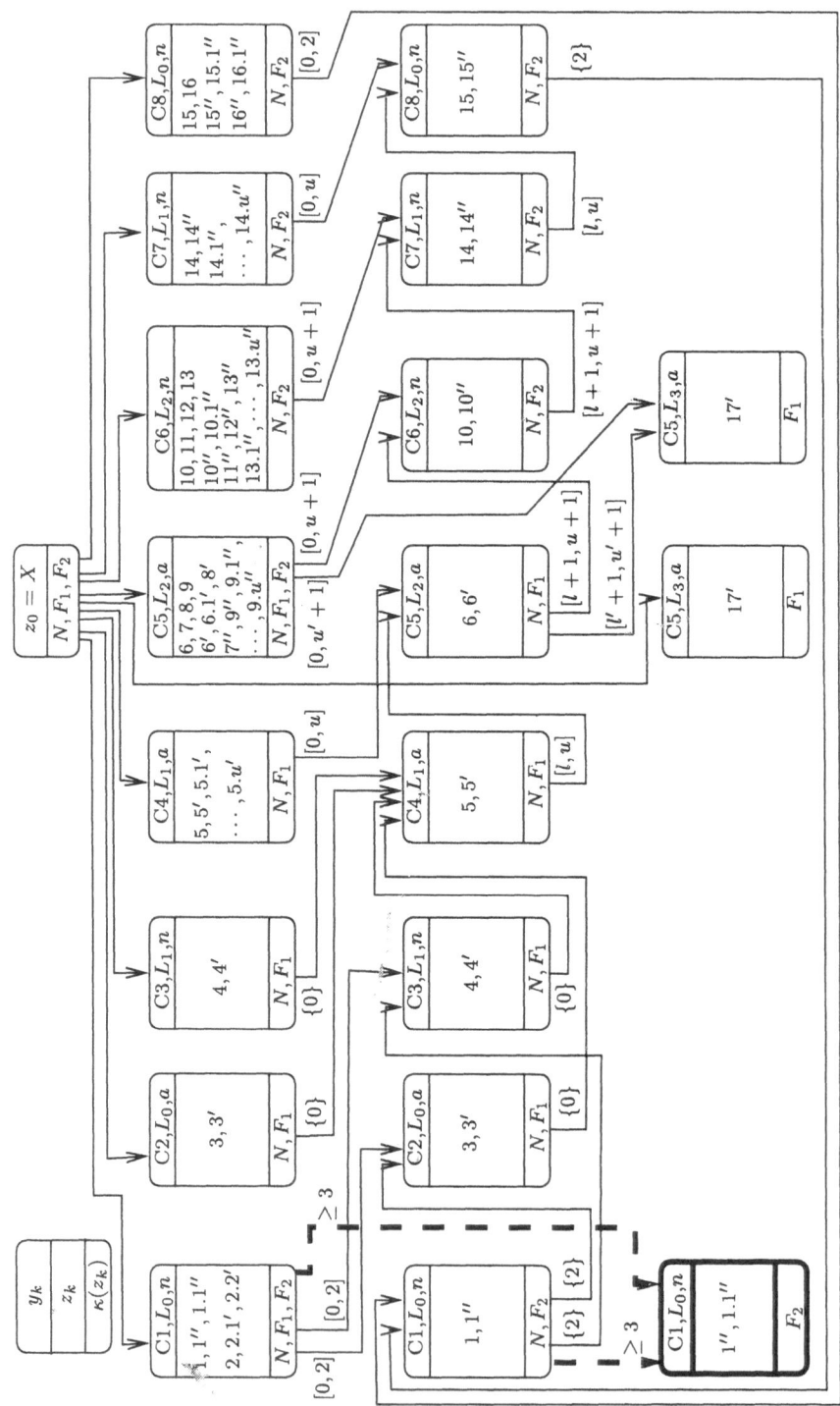

Fig. 10. Example 2. Diagnoser.

Table 5. Example 2. Timed reachability transition system.

State	Output-adjacent states {time}	State	Output-adjacent states {time}
1	3,3' {2} / 4,4' {2}	1''	-
2	3,3' {1} / 4,4' {1}	1.1''	-
3	5,5' {0}	7''	10'' {2,3}
4	5,5' {0}	9''	10'' {2,3}
5	6,6' {2,3}	9.1''	10'' {1,2}
6	10,10'' {3,4} / 17' {2,3}	9.2''	10'' {0,1}
7	10,10'' {2,3}	9.3''	10'' {0}
8	10 {2,3} / 17' {1,2}	10''	14'' {3,4}
9	10,10'' {2,3}	10.1''	14'' {2,3}
10	14,14'' {3,4}	11''	14'' {2,3}
11	14,14'' {2,3}	12''	14'' {2,3}
12	14,14'' {2,3}	13''	14'' {2,3}
13	14,14'' {2,3}	13.1''	14'' {1,2}
14	15,15'' {2,3}	13.2''	14'' {0,1}
15	1,1'' {2}	13.3''	14'' {0}
16	1,1'' {1}	14''	15'' {2,3}
2.1'	3' {1} / 4' {1}	14.1''	15'' {1,2}
2.2'	3' {0} / 4' {0}	14.2''	15'' {0,1}
3'	5' {0}	14.3''	15'' {0}
4'	5' {0}	15''	1'' {2}
5'	6' {2,3}	15.1''	1'' {1}
5.1'	6' {1,2}	16''	1'' {1}
5.2'	6' {0,1}	16.1''	1'' {0}
5.3'	6' {0}		
6'	17' {2,3}		
6.1'	17' {1,2}		
8'	17' {1,2}		
17'	-		

$\kappa(z_2) = \{F_1\}$; hence, the diagnoser will reach an F_1-certain state. According to the timed RTS, $(C5, L_3, a)$ can be generated within l' to $u' + 1$ ticks of the clock after y_1 is generated. Therefore y_2 can take any time between 0 to $u' + 1$ ticks to occur *following the initialization* of the diagnoser, i.e., $0 \le t_2 \le u' + 1$. The time interval $[0, u' + 1]$ is written on the transition from $\{6, 7, 8, 9, 6', 6.1', 8', 7'', 9'', 9.1'', \cdots, 9.u''\}$ to $\{17'\}$ to indicate this. Similarly, if instead of $(C5, L_3, a)$, $y_2 = (C6, L_2, n)$, then $z_2 = \{10, 10''\}$. This transition should occur in 0 to $u + 1$ ticks. After this, $y_3 = (C7, L_1, n)$ will be generated in $l + 1$ to $u + 1$ ticks $(l + 1 \le t_3 \le u + 1)$. The rest of Fig. 10 can be similarly interpreted.

The diagnoser provides estimates of the system's condition at the instants when new output symbols are generated. For diagnosis, we also need condition estimates between output symbols after every clock tick. In the following, we will discuss how for diagnosing permanent failures, these estimates can be replaced with *predictions* for the system's condition.

Let us assume that at some point an output symbol y is generated. Suppose z and $\kappa(z)$ are the estimates of the systems's state and condition following the generation of y (Fig. 11). Let $\mathrm{Rch}(z)$ denote the set of states of the timed

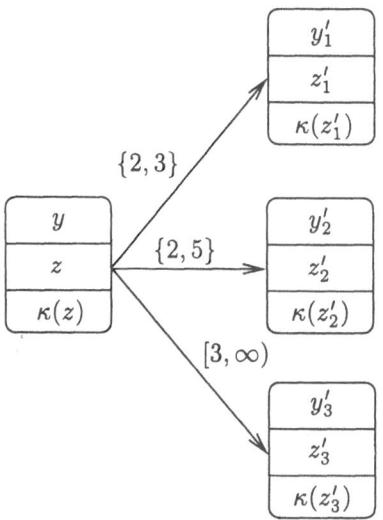

Fig. 11. Using predictions in fault diagnosis.

FSMA that have output y and are reachable from a state in z using a path along which the output is y. Also define

$$\mathrm{Lim}(z) := \{x \in X | \; x \in \mathrm{Rch}(z) \; \& \; (\delta(x,\sigma) = \emptyset \;\; \forall \sigma \in \Sigma)\} \cup$$
$$\{x \in X | \; x \in \mathrm{Rch}(z) \; \& \; (\exists\, x' \in X, \sigma \in \Sigma : \; |T(x,\sigma,x')| = \infty\} \cup$$
$$\{x \in X | \; x \text{ belongs to a cycle in } \mathrm{Rch}(z)\}.$$

$\mathrm{Lim}(z)$ is the set of states and cycles in $\mathrm{Rch}(z)$ in which the system can be trapped for an arbitrary long time without generating a new output symbol. As illustrated in Fig. 11, following y, either a new output symbol y_i' ($i \in \{1,2,3\}$) is generated after some time in which case $x \in z_i'$ at the time y_i' is observed or, no new output is generated in which case x must eventually enter $\mathrm{Lim}(z)$ after a finite time. For this reason, we refer to $\mathrm{Lim}(z) \cup (\cup_{i=1}^{3} z_i')$ as the *prediction* for the system's state at the time y was generated. If four ticks occur without the generation of any new output symbol, we can conclude that the next output symbol cannot be y_1'. Therefore the transition to z_1' can be ruled out and the prediction can be updated to $\mathrm{Lim}(z) \cup z_2' \cup z_3'$. More generally, we denote the prediction for the system's condition after t clock ticks following the generation of y, assuming the generation of no new output symbol, by

$\text{Pred}(z,t)$ and define it according to

$$\text{Pred}(z,t) := \text{Lim}(z) \cup (\cup_i \{z_i' \mid t \le \sup \hat{T}_i(z, z_i')\})$$

where $\hat{T}_i(z, z_i')$ denotes the set of transition times from z to z_i' in the diagnoser. In Fig. 11, $\hat{T}_1(z, z_1') = \{2, 3\}$.

Naturally, $\kappa(\text{Pred}(z,t))$ will be the prediction for the system's condition. Now suppose the failures are permanent and hence, once a failure occurs, it will remain permanently. Then we can expect to be able to use the predictions for fault detection and isolation. In fact, it can be shown [7] that after t ticks of clock following the occurrence of an observation y_k and before the (possible) generation of y_{k+1}, if the standard diagnoser is in an F_i-certain state, then $\text{Pred}(z_k, t)$ will also be F_i-certain (Recall that z_k is the state estimate provided by the diagnoser described in this section upon observing y_k). Therefore, our diagnoser is at least as fast as the standard diagnoser in detecting and isolating failures. In some cases, our diagnoser could even be faster than the standard diagnoser. For instance, if a failure F_2 is caused by another failure F_1, then our diagnoser may predict F_2 (when it detects F_1) even before F_2 occurs. Obviously the standard diagnoser cannot detect F_2 before it happens.

In summary, the diagnoser provides an estimate of the system's condition ($\kappa(z_k)$) every time a new output symbol is generated. Between consecutive output symbols, the predictions $\kappa(\text{Pred}(z_k, t))$ can be used for diagnosing permanent failures.

Going back to the diagnoser for the neutralization process (Fig. 10), we can easily verify that for $z = \{1, 1''\}$, $\text{Lim}(z) = \{1'', 1.1''\}$, and

$$\text{Pred}(z,t) = \{3, 3', 4, 4'\} \cup \{1'', 1.1''\} \text{ for } 0 \le t \le 2$$
$$\text{Pred}(z,t) = \{1'', 1.1''\} \qquad\qquad \text{ for } t \ge 3.$$

Therefore when the diagnoser is at $z = \{1, 1''\}$, if no new output symbol is generated for three ticks of the clock, then it can be concluded that F_2 has occurred. This deduction is shown on Fig. 10 by a transition (dashed line) from $z = \{1, 1''\}$ to $\text{Pred}(z,t) = \{1'', 1.1''\}$.

We define diagnosability in fault diagnosis with timing information as follows: A permanent failure mode F_i is *time-diagnosable* if there exists an integer $T \ge 0$ such that, following both the occurrence of the failure and initialization of the diagnoser, F_i can be detected and isolated (i.e., either z_k or $\text{Pred}(z_k, t)$ become F_i-certain) in at most T *clock ticks*. If the TDES is activity-loop-free, i.e., it contains no cycles of non-tick events, then the notions of diagnosability (sect. 2) and time-diagnosability become equivalent for the TDES [7]. In [7], necessary and sufficient conditions for time-diagnosability of failures are obtained. Also note that unlike the event-based version [3], in our definition of time-diagnosability, no assumption is made about the system's state or condition at the time the diagnoser is started.

For the neutralization process, the reader can verify that if V1 gets stuck-open, then in at most $u + u' + 2$ clock ticks, the output symbol $(C5, L_3, a)$ will be generated and the diagnoser will enter the F_1-certain state $z = \{17'\}$.

On the other hand, if V1 gets stuck-closed, then in at most $3u + 3$ ticks, the diagnoser will enter $z = \{1, 1''\}$ and $\texttt{Pred}(\{1, 1''\}, 3) = \{1'', 1.1''\}$ will indicate the occurrence of F_2. Hence, F_2 will be diagnosed in at most $3u + 6$ clock ticks. Therefore, both F_1 and F_2 are time-diagnosable. Note that without the timing information, F_2 would not have been diagnosable because once the tank is emptied and $(C1, L_0, n)$ is generated, the system stops generating new output symbols and the estimate of the system's condition remains ambiguous.

The fact that in our approach, no update of the estimate of the systems's state is required at clock ticks results in significant reduction in on-line computing power requirement. In many cases, the diagnoser will also have fewer states than the standard diagnoser. For example, in the neutralization process, filling the tank takes a lot longer than opening a valve (assumed here to take 1 clock tick). Therefore u and u' are considerably larger than 1. If we assume $u = 200$ and $u' = 100$, then the TDES will have about 1700 states. As a result, while both the standard diagnoser and the diagnoser based on the methodology of [3] will have at least a few hundred states, our diagnoser (Fig. 10) has only 19 states and, except for some of the transition time sets $(\hat{\mathcal{T}}(., .))$, the structure of our diagnoser does not depend on the parameters l, u, l' and u'. In general, a small clock period is necessary for accurate time keeping in a TDES. This results in a large state set for the TDES. In these cases, our approach can lead to significant reduction in the size of the diagnoser and in the on-line computing power requirement. This improvement is obtained at the expense of more off-line design calculations, in particular, the computation of the transition time sets $\mathcal{T}$ and $\tilde{\mathcal{T}}$. For further details, the reader is referred to [7].

4 Conclusion

In this paper, we have proposed a state-based framework for fault diagnosis in finite-state automata and timed discrete-event systems. In this approach, the system and the diagnoser (the fault detection system) do not have to be started at the same time and no information about the state or even the condition of the system before the initialization of the diagnoser is required. Furthermore, any problem in an event-based framework can be recast as a problem in our state-based framework. We have shown how we can reduce model complexity using a polynomial time algorithm. We have also presented a new approach to incorporate timing information which does not require estimate updates at clock ticks, thereby significantly reducing on-line computing power requirements and, in many cases, the size of the diagnoser.

We have focused on the main ideas in this paper and have illustrated them with examples. A more detailed account of our work, including necessary and sufficient conditions for diagnosability and extension to hybrid systems, can be found in [7].

References

1. B.A. Brandin and W.M. Wonham, "Supervisory control of timed discrete-event systems," *IEEE Trans. Automat. Contr.*, vol. 39, no. 2, pp. 329–342, 1994.

2. S. Chand, "Discrete-event based monitoring and diagnosis of manufacturing processes," in *Proc. Amer. Contr. Conf.*, San Francisco, CA, June 1993, pp. 1508–1512.

3. Y. Chen and G. Provan, "Modeling and diagnosis of timed discrete event systems — A factory automation example," in *Proc. Amer. Contr. Conf.*, Albuquerque, NM, June 1997, pp. 31–36.

4. T.Y.L. Chun, "Diagnostic supervisory control: A DES approach," Master's thesis, Dept. of ECE, University of Toronto, Canada, Aug. 1996.

5. W. Hamscher, L. Console and J. de Kleer, Eds., *Readings in Model-Based Diagnosis.* San Mateo, CA: Morgan Kaufmann, 1992.

6. S. Hashtrudi Zad, R.H. Kwong and W.M. Wonham, "Fault diagnosis in discrete-event systems: Framework and model reduction," to be presented at the *37th IEEE Conf. Decision Contr.*, Tampa, FL, USA.

7. S. Hashtrudi Zad, Ph.D. thesis, Dept. of ECE, University of Toronto, Canada, in preparation.

8. D. Lee and M. Yannakakis, "Principles and methods of testing finite state machines — A survey," *Proc. IEEE*, vol. 84, no. 8, pp. 1090–1123, 1996.

9. W.S. Lee, D.L. Grosh, F.A. Tillman and C.H. Lie, "Fault tree analysis, methods, and applications — A review," *IEEE Trans. Reliability*, vol. R-34, no. 3, pp. 194–203, 1985.

10. F.P. Lees, *Loss Prevention in the Process Industries, Volume 1.* London: Butterworths, 1980.

11. N.G. Leveson, *Safeware: System Safety and Computers.* Reading, Mass.: Addison-Wesley, 1995.

12. F. Lin, "Diagnosability of discrete event systems and its applications," *Discrete Event Dynamic Systems*, vol. 4, pp. 197–212, 1994.

13. R. Paige and R.E. Tarjan, "Three partition refinement algorithms," *SIAM J. Comput.*, vol. 16, pp. 973–989, 1987.

14. D.N. Pandalai and L.E. Holloway, "Template languages for fault monitoring of single-instance and multiple-instance discrete event processes," in *Proc. 36th Conf. Decision Contr.*, San Diego, USA, Dec. 1997, pp. 4619–4625.

15. T. Ruokonen, Ed., *Proc. of IFAC Symp. SAFEPROCESS'94*, Espoo, Finland, 1994.

16. M. Sampath, R. Sengupta, S. Lafortune, K. Sinnamohideen and D. Teneketzis, "Diagnosability of discrete-event systems," *IEEE Trans. Automat. Contr.*, vol. 40, no. 9, pp. 1555–1575, 1995.

17. M. Sampath, R. Sengupta, S. Lafortune, K. Sinnamohideen and D. Teneketzis, "Failure diagnosis using discrete-event models," *IEEE Trans. Contr. Syst. Technology*, vol. 4, no. 2, pp. 105–124, 1996.

18. M. Sampath, S. Lafortune and D. Teneketzis, "Active diagnosis of discrete-event systems," in *Proc. 36th Conf. Decision Contr.*, San Diego, USA, Dec. 1997, pp. 2976–2983.

19. N. Viswanadham and T.L. Johnson, "Fault detection and diagnosis of automated manufacturing systems," in *Proc. 27th Conf. Decision Contr.*, Austin, USA, Dec. 1988, pp. 2301–2306.

On The Stability of Control Systems Having Clegg Integrators

Hanzhong Hu[1], Yuan Zheng[2], Christopher V. Hollot[3], and Yossi Chait[4]

[1] Security Dynamics, 20 Crosby Drive, Beford, MA 01730 USA
   Email: jhu@securid.com
[2] NeoMagic Corporation, Santa Clara, CA 95054 USA
   Email: yzheng@neomagic.com
[3] ECE Department, University of Massachusetts, Amherst, MA 01003 USA
   Email: hollot@ecs.umass.edu
[4] MIE Department, University of Massachusetts, Amherst, MA 01003 USA
   Email: chait@ecs.umass.edu

Abstract. In this paper, we consider control systems utilizing so-called Clegg integrators. A Clegg integrator is a linear integrator with reset mechanism whose describing function $\frac{1.62}{\omega} \angle -38.1°$ has magnitude slope equivalent to that of a linear integrator, but with 52° less of phase lag. The potential advantages of using Clegg integrators have been demonstrated in the literature via both simulations and experiments. However, except for describing function analysis, stability criteria specifically tailored to these feedback systems is missing. This paper addresses the internal stability of such control systems and provides stability conditions when the plant is second-order and preliminary results for higher-order cases.

1 Introduction

A linear integrator provides infinite gain at low frequencies enabling feedback systems to provide zero steady-state errors in response to constant exogenous signals. This benefit, however, comes at the expense of $-90°$ phase shift. Indeed, for a stable feedback system, the integrator's phase shift together with Bode's gain-phase relation limits the rate at which the open-loop transfer function transitions from high to low gain. To reduce sensor noise amplification and instability due to unmodeled high frequency dynamics, it is desirable to attenuate the open-loop gain as much as possible at high frequencies. Thus, the control engineer is faced with a tradeoff between a linear integrator's low-frequency benefits and its undesirable impact on high-frequency loop gain.

An effort to favorably improve this tradeoff prompted the development of the Clegg integrator (see [1]) which is a linear integrator which resets to zero whenever its input crosses zero. Its describing function $\frac{1.62}{\omega} \angle -38.1°$ has magnitude slope similar to the frequency response of its linear counterpart but with significantly less phase lag. Consequently, it seems possible that the Clegg integrator can provide solutions to specialized control problems that cannot be solved with linear integrators. This possibility is supported by the design methodologies, simulations and experiments given in [2]–[4]. However, this work did not address the issue of closed-loop stability. Moreover, except for the sufficient conditions for oscillation and non-oscillation given in the general context of describing functions (e.g., see [5]), it appears that no stability analysis for feedback systems utilizing Clegg integrators has been undertaken. With this motivation, we begin a study of this specific stability problem and report some first results which help characterize the internal stability of a feedback interconnection of a Clegg integrator with a second-order plant.

Before addressing this situation in detail, we first analyze the behavior of a Clegg integrator in feedback with a first-order plant. Therefore, consider the system consisting of a negative feedback interconnection of a linear time-invariant plant with transfer function $G(s)$ (output y) and a Clegg integrator (input x_c). To model this reset element, let t_i be an instance when $y(t_i) = 0$. Then, the output of a Clegg integrator is described by:

$$x_c(t) = \begin{cases} -\int_{t_i}^{t} y(t)dt, & t \neq t_i \\ 0, & t = t_i. \end{cases} \tag{1}$$

Let $G(s)$ be the first-order plant

$$G(s) = \frac{k}{s + \alpha} \tag{2}$$

with $\alpha, k > 0$. Notice that $G(s)$ is stabilized by a linear integrator. To study the internal stability of this feedback system, consider non-zero initial conditions and assume the first reset-time of $y(t)$ occurs at t_1. Consequently,

$$y(t_1) = 0; \quad x_c(t_1) = 0$$

which implies $y(t) = 0$ and $x_c(t) = 0$ for all $t > t_1$. On the other hand, if $y(t) \neq 0$ for all $t \geq 0$, then the Clegg integrator never resets and behaves like a linear integrator which stabilizes $G(s)$. In either case, the closed-loop system with Clegg integrator is internally stable and for first-order plants the Clegg integrator resets at most once and behaves linearly thereafter. Higher-order plants exhibit more complex reset behavior, and the remainder of this paper is geared to giving necessary and sufficient conditions for the internal stability of Clegg control systems with second-order plants. These results are achieved by showing that the reset times of $y(t)$ are periodically spaced, reducing the stability problem to a linear, shift-invariant discrete-time problem.

2   Second-Order Plants

Assume that the strictly-proper plant $G(s) = c(sI - A)^{-1}b$ is a second-order transfer function. Then, the closed-loop system has a state-space representation:

$$\dot{x}(t) = Ax(t) + bx_c(t), \quad t \geq 0;$$
$$y(t) = cx(t);$$
$$\dot{x}_c(t) = -y(t), \quad t \neq t_i;$$
$$x_c(t_i) = 0 \tag{3}$$

where $A \in \mathbb{R}^{2 \times 2}$, $b \in \mathbb{R}^{2 \times 1}$, $c \in \mathbb{R}^{1 \times 2}$, $x(t)$ is the plant state and $x_c(t)$ is the Clegg integrator's state. Define the *reset times* t_i as those instances for which $y(t_i) = 0$ and $x(t_i) \neq 0$.[1] Collect the reset times in the ordered set $\{t_i : t_i < t_{i+1},\ i = 1, 2, \ldots, \}$.[2] Between successive reset times t_i and t_{i+1}, the closed-loop plant state behaves as the linear time-invariant system

$$x(t) = P(t - t_i)x(t_i), \quad t \in (t_i, t_{i+1}) \tag{4}$$

where

$$P(t) = [I \quad 0]\, e^{A_{cl} t} \begin{bmatrix} I \\ 0 \end{bmatrix} \tag{5}$$

and

$$A_{cl} = \begin{bmatrix} A & b \\ -c & 0 \end{bmatrix}.$$

Notice that A_{cl} is the state matrix for the linear closed-loop system involving plant $G(s)$ and a unity-gain linear integrator. By definition, a *reset time* t_i is characterized by $cx(t_i) = 0$. Thus, from (4), the reset times t_i must satisfy:

$$cx(t_1) = 0$$
$$cP(\tau_1)x(t_1) = 0$$
$$cP(\tau_2)P(\tau_1)x(t_1) = 0$$
$$\vdots \tag{6}$$

where $\tau_i = t_{i+1} - t_i$. We say that the reset times t_i are *periodically-spaced* whenever (6) is satisfied with $\tau_i = \tau_{i+1}$ for $i = 1, 2, \ldots$. The next lemma further characterizes reset times for second-order plants.

Lemma 1. *Suppose the closed-loop system (3) consisting of second-order plant $G(s)$ and Clegg integrator has at least one reset time t_1. Then, the reset times t_i are characterized as follows:*

1. *If c is not a left eigenvector of $P(\tau)$ for all $\tau > 0$, there is exactly one reset time t_1.*
2. *If c is a left eigenvector of $P(\tau)$ for some $\tau > 0$, then the reset times are infinite in number and periodically-spaced. Moreover, the smallest τ satisfying this eigenvector condition is the period of these reset times.*

---

[1] The condition $x(t_i) \neq 0$ is included to avoid calling all $t > t_i$ reset times. That is, if $x(t_i) = 0$, then $x(t) = 0$ for all $t > t_i$ since $(x, x_c) = (0, 0)$ is an equilibrium point of (3).

[2] To allow for this ordering, we assume that the switching surface $y(t) = 0$ does not possess a sliding mode. Consequently, there exists an $\epsilon > 0$ such that $t_{i+1} - t_i > \epsilon$ for all i. Also, $t_i \to \infty$ if $i \to \infty$.

Proof. 1. Suppose c is not an eigenvector of $P(\tau)$ for all $\tau > 0$. Then,

$$\det \begin{bmatrix} c \\ cP(\tau) \end{bmatrix} \neq 0 \qquad (7)$$

for all $\tau > 0$. Proceeding by contradiction, suppose there are two reset times. t_1 and t_2. Then, $cx(t_1) = 0$ and $cP(t_2 - t_1)x(t_1) = 0$ which together with both (7) and $P(t) \in \mathbb{R}^{2 \times 2}$ implies $x(t_1) = 0$. Hence, t_1 is not a reset time which is a contradiction.

2. Let $\tau > 0$ be the smallest number for which c is a left eignevector of $P(\tau)$. We will show that each of the intervals τ_i are equal to τ and hence prove that the reset times t_i are periodically-spaced. By assumption, there exists a non-trivial reset time so that

$$cx(t_1) = 0 \qquad (8)$$

for some t_1 and $x(t_1) \neq 0$. The first reset time interval $\tau_1 > 0$ is the smallest number for which

$$cP(\tau_1)x(t_1) = 0. \qquad (9)$$

Taken together, (8) and (9) are the same as

$$\begin{bmatrix} c \\ cP(\tau_1) \end{bmatrix} x(t_1) = 0.$$

This is equivalent to c being a left eigenvector of $P(\tau_1)$ which proves $\tau_1 = \tau$. To show $\tau_2 = \tau$, notice that the second reset time interval $\tau_2 > 0$ is the smallest number for which

$$cP(\tau_2)P(\tau_1)x(t_1) = 0. \qquad (10)$$

In this case, (8) – (10) become

$$\begin{bmatrix} c \\ cP(\tau_1) \\ cP(\tau_2)P(\tau_1) \end{bmatrix} x(t_1) = 0. \qquad (11)$$

Since $c \in \mathbb{R}^2$ and $P(\tau_1)$ is nonsingular, (11) reduces to

$$\begin{bmatrix} c \\ cP(\tau_2) \end{bmatrix} x(t_2) = 0.$$

This is again equivalent to c being a left eigenvector of $P(\tau_2)$ which proves $\tau_2 = \tau$. Continuing this argument one can show that $\tau_i = \tau$ for $i = 3, 4, \ldots$.

Finally, since $x(t_1) \neq 0$ and $e^{A_{cl}\tau}$ is nonsingular, then

$$x(t_{i+1}) = P(\tau)x(t_i) \neq 0$$

for $i = 1, 2, \ldots$. Thus, the reset times are infinite in number. □

Remark 1. Lemma 1 states that if the Clegg integrator resets, then one of two things must occur: either $x_c(t)$ has an infinite number of reset times or exactly one reset time. In the latter case, (3) is stable if $G(s)$ is stabilized by a linear integrator. In the former case, when there is an infinite number of periodic reset times, stability may be deduced from the state's behavior at these periodic instances. Our main result below exploits this periodicity and gives a stability condition in terms of a shift-invariant, discrete-time system.

Before giving our main result, we formally define internal stability for (3) where we denote the closed-loop state as

$$z = \begin{bmatrix} x \\ x_c \end{bmatrix}.$$

Definition 1. The closed-loop system (3) is *internally uniformly exponentially stable* (or *internally stable* for short) if there exist finite positive constants γ and λ such that for any t_0 and z_0, the corresponding solution satisfies

$$\|z(t)\| \leq \gamma e^{-\lambda(t-t_0)}\|z_0\|, \quad t \geq t_0. \tag{12}$$

Also, if (12) holds for some z_0, we say the solution $z(t)$ (corresponding to z_0) *exponentially converges* to the origin.

In the following, $\rho[P(\tau)]$ denotes spectral radius of matrix $P(\tau)$.

Theorem 1. *Assume that the strictly-proper second-order plant $G(s)$ is stabilized by a unity-gain linear integrator; i.e., A_{cl} has eigenvalues only in the left-half of the complex plane. Then, the closed-loop system (3), consisting of plant $G(s)$ and Clegg integrator, is internally stable if and only if $\rho[P(\tau)] < 1$ where $\tau > 0$ is the smallest number for which c is a left eigenvector of $P(\tau)$.*

Proof. (Sufficiency) First consider the vacuous case when c is not a left eigenvector of $P(\tau)$ for all $\tau > 0$. Then, from Lemma 1, the Clegg integrator resets no more than once. Hence, (3) eventually behaves as a feedback interconnection of $G(s)$ and a linear integrator, which, by assumption, is internally stable. Now consider the case when the $\rho[P(\tau)] < 1$ and c is a left eigenvector of $P(\tau)$. To avoid the trivial situation, assume the Clegg integrator resets at least once. This implies, from Lemma 1, that the reset times are periodic and infinite in number. Consequently, using (4), the plant state satisfies the equation

$$x(t_{i+1}) = P(\tau)x(t_i), \quad i = 1, 2, \dots. \tag{13}$$

Because $\rho[P(\tau)] < 1$, this discrete-time system is internally stable. It is now straightforward to show the internal stability of (3). Indeed, between successive reset times

$$z(t) = [I \quad 0]e^{A_{cl}t}z(t_i), \quad t \in (t_i, t_{i+1})$$

so that

$$\|z(t)\| \leq \delta\|z(t_i)\|, \quad t \in (t_i, t_{i+1}) \tag{14}$$

where
$$\delta = \max_{t\in[0,\tau]} \|[I \quad 0]e^{A_{cl}t}\|.$$

Since $x_c(t_i) = 0$ for $i = 1, 2, \ldots$ then from (13),
$$\|z(t_i)\| \le \bar{\rho}^i\|z(0)\|, \quad i = 1, 2, \ldots$$

where $\bar{\rho} := \rho[P(\tau)]$. From (14),
$$\|z(t)\| \le \delta e^{\frac{\ln(\bar{\rho})}{\tau}t}\|z(0)\|; \quad t \ge 0.$$

Setting $\gamma = \delta$ and $\lambda = -\ln(\bar{\rho})/\tau$ shows (12) and the internal stability of (3).

(Necessity) Proceeding by contradiction, suppose $\rho[P(\tau)] \ge 1$. Given any time $t_1 > 0$, there exists an initial condition z_0 such that the Clegg integrator resets at $t = t_1$; that is, $cx(t_1) = 0$. From Lemma 1, the reset times are periodic and infinite. Consequently, from (4), the plant state satisfies (13). Because $\rho[P(\tau)] \ge 1$, this discrete-time system is internally unstable which implies that (3) is internally stable. This proves the theorem.   □

When $G(s)$ is destabilized by a linear integrator; i.e., when A_{cl} has some eigenvalues in the closed right-half complex plane, we have the following characterization of solutions to (3).

Theorem 2. *Assume that the strictly-proper second-order plant $G(s)$ is destabilized by a unity-gain linear integrator. As in the previous theorem, let $\tau > 0$ be the smallest number for which c is a left eigenvector of $P(\tau)$. Then, the following hold:*

1. *If the response of (3) due to initial condition z_0 does not result in the Clegg integrator resetting, then the response $z(t)$ does not converge to the origin.*
2. *If the response of (3) due to initial condition z_0 results in the Clegg integrator resetting at least once, then the Clegg integrator will continue to reset with period τ and the response $z(t)$ will exponentially converge to the origin if and only if $\rho[P(\tau)] < 1$.*

Proof. The proof of the first situation is straightforward. Therefore, consider the second case when the Clegg integrator resets at least once. From Lemma 1, the reset times t_i are infinite and periodic with period τ. Consequently, from (4), the plant state satisfies (13). If $\rho[P(\tau)] < 1$, then (13) is internally stable. As shown in the proof of Theorem 1, this is enough to show that the response $z(t)$ in (3) exponentially converges to the origin. If $\rho[P(\tau)] \ge 1$, the discrete-time system (13) is internally unstable which implies that (3) is also internally unstable. This proves the theorem.   □

3   Examples

Example 1. In our first example, we illustrate Theorem 1. Consider the feedback system (3) with
$$G(s) = \frac{(3+\beta)s+1}{s^2 + 3s - \beta}$$

and β a free parameter. For all β, $G(s)$ is stabilized by a linear integrator. Similarly, for all β, $\tau = 2$ is the smallest positive number for which c is a left eigenvector of $P(\tau)$. Thus, from Lemma 1, the Clegg integrator resets every 2 seconds, independent of β. The spectral radius of $P(2)$ is $\max\{|5+2\beta|e^{-2}, e^{-2}\}$; hence, if β satisfies

$$|(5+2\beta)e^{-2}| < 1,$$

then $\rho[P(2)] < 1$ and the system is stable. On the other hand, if

$$|(5+2\beta)e^{-2}| \geq 1,$$

then the system is unstable. Therefore, this Clegg control system is internally stable if and only if $\beta \in (-6.195, 1.195)$. It is interesting to note that describing function analysis predicts the liklihood of a limit cycle when $\beta = -7.1$. In Figure 1 we plot the results of a SIMULINK simulation of the Clegg control system for $\beta = -5$ and initial conditions $y(0) = \dot{y}(0) = 1$ and $x_c(0) = 0$.

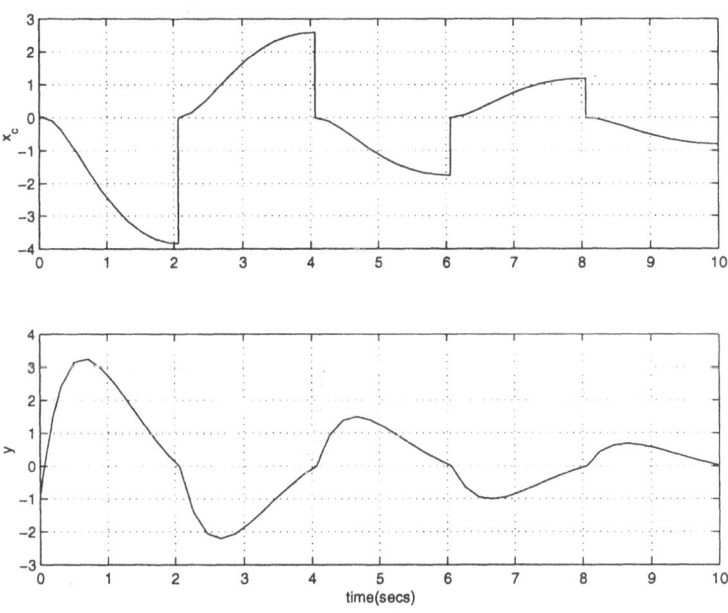

Fig. 1. SIMULINK simulation for Example 1 where $\beta = -5$ and $G(s) = \frac{-2s+1}{s^2+3s+5}$. For initial conditions $y(0) = \dot{y}(0) = 1$ and $x_c(0) = 0$, the top figure shows the Clegg integrator's output x_c while the bottom displays the plant's response y. Both the reset period $\tau = 2$ and stable behavior are verified.

Example 2. Now consider Theorem 2 and the transfer function plant

$$G(s) = \frac{3.1s + 1}{s^2 + .3322s - .1}.$$

This plant is not stabilized by a unity-gain linear integrator. However, c is a left eigenvector of $P(1.6407)$ and $\rho[P(\tau)] = .95$. Thus, using the second case in Theorem 2, there exist initial conditions z_0 for which the response of (3) is stable. For example, consider a state-space representation of $G(s)$ having data:

$$A = \begin{bmatrix} 0 & 1 \\ 0.1 & -0.3322 \end{bmatrix}; \quad b = \begin{bmatrix} 0 \\ 1 \end{bmatrix}; \quad c = \begin{bmatrix} 1 & 3.1 \end{bmatrix}.$$

Then, for the initial condition $z_0 = [-.95 \ .3 \ .1]$, the Clegg integrator resets periodically with period 1.6407 seconds and $z(t)$ exponentially converges to the origin. On the other hand, this same initial condition produces an unstable response in the linear feedback system involving $G(s) = c(sI - A)^{-1}b$ and a unity-gain integrator.

4 Higher-Order Plants

As shown in Lemma 1, the Clegg integrator resets periodically when in feedback with a second-order plant. When connected to higher-order plants, the reset patterns are more complex. At this stage of research, we have not characterized these patterns. However, for an nth-order plant, we can predict when the Clegg integrator will reset $n - 1$ times at most. In these cases, the internal stability of the closed-loop system (3) depends solely on whether the plant is stabilized by a linear integrator. To state this result, consider the set-up in (3) where now $A \in \mathbb{R}^{n \times n}$, $b \in \mathbb{R}^{n \times 1}$ and $c \in \mathbb{R}^{1 \times n}$. With $P(t)$ as in (4) and (5), define

$$\mathcal{T}(\tau_1, \tau_2, \ldots, \tau_{n-1}) = \begin{bmatrix} c \\ cP(\tau_1) \\ cP(\tau_2)P(\tau_1) \\ \vdots \\ cP(\tau_{n-1}) \ldots P(\tau_1) \end{bmatrix}.$$

Theorem 3. *If $\mathcal{T}(\tau_1, \tau_2, \ldots, \tau_{n-1})$ is nonsingular for all $\tau_i > 0$, $i = 1, 2, \ldots, n - 1$, then the Clegg integrator resets $n - 1$ times at most and (3) is internally stable if and only if the A_{cl} has eigenvalues only in the open left-half complex plane.*

Proof. We first show that $y(t)$ has $n - 1$ reset times at most. Indeed, suppose there are more. Then, the first n equations of (6) are satisfied for some τ_i, $i = 1, 2, \ldots, n - 1$. Since $\mathcal{T}(\tau_1, \tau_2, \ldots, \tau_{n-1})$ is nonsingular, $x(t_1) = 0$. Thus, t_1 is not a reset time which is a contradiction. Therefore, the largest number of reset times is $n - 1$ and the Clegg eventually behaves as a linear integrator. Hence, (3) is internally stable if and only if $G(s)$ is stabilized by a unity-gain linear integrator, or equivalently, that A_{cl} has eigenvalues only in the open left-half complex plane. □

5   Conclusion

In this paper, we have studied the internal stability of a feedback connection between a Clegg integrator and a second-order plant transfer function. We show that if the Clegg integrator continually resets, then it must reset periodically. The periodic rate can be determined a priori from the plant description. As a consequence, internal stability can be deduced from an implicit shift-invariant discrete-time system. Also, these results show that a plant stabilized by a linear integrator can either remain stabilized or be destabilized by a Clegg integrator. Conversely, a plant which is destabilized by a linear integrator can exhibit a stable initial condition response even though this same initial condition excites unstable modes in the related linear feedback system. For feedback systems involving higher-order plants, we give conditions for which the Clegg integrator resets only a finite number of times. In addition to studying the more complex reset patterns for these high-order systems, future research will focus on input-output stability. In this case, the reset behavior is input dependent.

Acknowledgement

We would like to acknowledge Mr. Orhan Beker's help in pointing out and suggesting fixes to errors in earlier versions of the examples.

References

1. J. C. Clegg, "A nonlinear integrator for servomechanism," *Trans A.I.E.E.* Part II, 77, pp. 41–42, 1958.
2. I. M. Horowitz and P. Rosenbaum, "Nonlinear design for cost of feedback reduction in systems with large parameter uncertainty," *International Journal of Control*, vol. 21, pp. 977–1001, 1975.
3. Y. Zheng, "Theory and practical considerations in reset control," Ph.D. Dessertation, University of Massachusetts, Amherst, 1998.
4. Y. Zheng, Y. Chait, C. V. Hollot, M. Steinbuch and M. Norg, "Experimental demonstration of reset control design," University of Massachusetts Technical Report, also submitted for publication, September, 1998.
5. H. K. Khalil, *Nonlinear Systems*, Macmillan, 1992.

Constructive Solutions to a Decentralized Team Problem with Application in Flow Control for Networks*

Orhan Çağrı İmer and Tamer Başar

Department of Electrical and Computer Engineering, Coordinated Science Laboratory, University of Illinois, 1308 W. Main Street, Urbana, Illinois 61801-2307, USA
Email: imer, tbasar@decision.csl.uiuc.edu

Abstract. The paper considers a decentralized stochastic team problem with a *partially-nested* information pattern, that arises in the context of flow control. Basically, we consider a network with a number of users having access to differently delayed versions of the same information, with each one deciding on his own rate of transmission, but participating in a common cost quantifying the outcome of their joint actions. This leads to a Linear-Quadratic-Gaussian (LQG) team problem, with partially nested information. We study the derivation of the optimal solution in a two user network and show that the solution exists in both finite and infinite-horizon cases. The controller, which turns out to be certainty-equivalent, is constructed recursively using a dynamic programming type approach. We also present an algorithm to construct the optimal solution for the most general case with multiple (more than two) users. Finally, we present various simulation results to illustrate the performance of the optimal controller under different scenarios.

1 Introduction

It is a well-known and well-acknowledged fact that in stochastic teams the information pattern plays an important role in the existence as well as the derivation of optimal team solutions. The simplest information pattern one can think of is the so-called *classical* one, in which all members of a team receive the same information and have perfect recall. The next level is the *partially nested* one, which has the property that if a team member's information depends on the control variable of some other member, then the former has access to all the information accessible to the latter. It is well-known [7] that a stochastic team problem defined on a finite horizon and with partially nested information structure can be transformed to a static team problem (albeit of a much higher dimension), with the equivalence being in the sense that the solution of one can be obtained from the solution of the other. This equivalence readily leads to existence results, such as the one of the partially nested linear-quadratic-Gaussian team, where the solution exists, is unique, and is

---

* Research that led to this paper was supported in part by the AFOSR MURI Grant AF DC 5-36128. Corresponding author: T. Başar, tel: 217-333-3607, fax: 217-244-1653.

affine in the available information [7] – a result that follows from an existence and uniqueness result for quadratic Gaussian teams [9].

The equivalence alluded to above does not lead, however, to any constructive methods for obtaining the team-optimal solution for a general partially nested information structure, nor to closed-form solutions[1]. It also does not say much about the solution of infinite-horizon team problems with partially nested information. It is the latter class of problems that we will be addressing in this paper, for a specific model motivated by an application involving congestion control in communication networks.

Hence, we consider here a network of a number of users, viewed as members of a team, having access to differently delayed versions of the same information. Each user controls his own rate of transmission and the state of the network is described by a difference equation. The expected cost to be minimized is quadratic in state and the decision variables. The motivation behind this formulation is justified in a real network with a bottleneck node that plays a major role in determining the performance of a number of users. Although in a real network environment, users may be interconnected in several ways, the single bottleneck node assumption admits theoretical as well as experimental justification [6]. The available service rate is modeled as an autoregressive (AR) process driven by a white noise process, as in [4]. The state of the network is nothing but the queue length at the bottleneck node. To decide on the rate of transmission of each user, we assume that both queue length and total service rate information is available to the node, but the transmission of these decisions taken by a node to the sources (users) incur different (propagation) delays, depending on the distance between the node and each user. Hence, even though it may appear that this problem is a centralized one with the decisions made only by the node, but with *action delays*, it can in fact be shown that it is equivalent to one where the decisions are made at the sources (by the users), but with information delay (see, [2]).

The information pattern here is hence partially nested, since each user simply has a subset of the information available to the users with smaller delays. This special structure with a different information delay for each user poses a fundamental difficulty in the actual computation of the solution. One way to circumvent this difficulty is first to solve the problem by ignoring the delays. In this case, it can be shown that the problem can be reduced to a standard discrete-time linear-quadratic-regulator problem, which is known to admit a unique linear solution. Then, one can incorporate the delays into the controllers, using the certainty equivalence approach. An earlier paper [2] has employed this method to obtain the solution to the flow control problem and has presented two sub-optimal certainty-equivalent controllers, called *Controller 1* and *Controller 2*. The controller constructed here being optimal, outperforms both of these controllers, as the simulation results also corrob-

---

[1] If the partially nested information is of a special type, such as *one-step-delay information sharing* pattern, then a closed-form solution can be obtained by recursive decomposition [10], [5].

orate; it is however more complex than either Controller 1 or Controller 2 of [2].

As can be deduced from the preceding discussion, the problem actually admits multiple certainty-equivalent solutions, all of which however not being optimal. Here we obtain the optimal certainty-equivalent solution, and show that this is actually globally team optimal. The derivation of the team-optimal controller has been carried out rigorously, and the controller is expressed in closed form, which is linear in the two user case. For the general multiple-user case, we present an algorithm for the computation of the optimal solution, and present various simulation results to support and illustrate the approach taken.

This paper is organized as follows. In Sect. 2, we introduce and motivate the mathematical model. The derivation of the optimal controller for the two user case is presented in Sect. 3, and Sect. 4 extends this derivation to the multiple user case. Section 5 is devoted to presentation of the simulation results. The paper ends with the concluding remarks of Sect. 6.

2   Mathematical Model

The mathematical formulation of the problem follows along the lines of [1,2]. We consider a set $\mathcal{M} = \{1, \dots, M\}$ of users that share a common bottleneck node in a network. In our model, the time unit corresponds to the round trip delay, which is the time it takes for a packet to reach its destination and come back. Let q_n denote the queue length at the bottleneck link, and μ_n denote the effective service rate available in that link at the beginning of the nth time slot. Let $r_{m,n}$ denote the effective rate of user $m \in \mathcal{M}$ during the nth time slot. This rate may actually be the outcome of an action taken by the user m several time steps earlier. In our formulation, we will not recognize this delay explicitly, but instead include a delay factor in the information available to each user for the construction of transmission rates for that user. As shown in [2], these two formulations are equivalent.

Now, in terms of the notation introduced, the queue length evolves according to

$$q_{n+1} = q_n + \sum_{m=1}^{M} r_{m,n} - \mu_n. \tag{1}$$

The above equation corresponds to a linearized version of the actual queue dynamics. Specifically, we ignore the fact that the queue length is restricted to be positive. Also, we assume no upper bound on the queue length. Simulations in [2], [3] show that these are valid assumptions. The service rate μ_n available to the M sources may change over time in an unpredictable way. We model this by a p-dimensional stable AR process:

$$\mu_n = \mu + \xi_n, \quad \xi_n = \sum_{i=1}^{p} \alpha_i \xi_{n-i} + \phi_{n-1}$$

where μ is the constant nominal service rate (known to all sources), and α_i, $i = 1, \ldots, p$, are known parameters. $\{\phi_n\}_{n \geq 1}$ is a zero mean $i.i.d$ sequence with finite variance σ_ϕ^2.

The objective function to be minimized (collectively by all M users) is

$$J = \limsup_{N \to \infty} \frac{1}{N} E \left\{ \sum_{n=1}^{N} \left[(q_n - Q)^2 + \sum_{m=1}^{M} c_m^2 \left(r_{m,n} - a_m \mu_n \right)^2 \right] \right\}$$

where Q is the target queue length, $\sum_{m=1}^{M} a_m = 1$, and c_m's are positive constants. The first additive term above represents a penalty for deviating from a desirable queue length. The second additive term is a measure of the quality with which the input rate for each user tracks a given fraction of the available service rate, where c_m's are weighting factors that serve to prioritize relative importance of these. For example, if we desire *fair* sharing of the available bandwidth, we would choose

$$a_1 = a_2 = \ldots = a_M = \frac{1}{M}$$

assuming that everything else is symmetric for the sources.

The information available to user m at time n is I_{n-D_m}, where

$$I_n := \{q_1, q_2, \ldots, q_n; \mu_1, \mu_2, \ldots, \mu_n\}$$

and D_m's denote delays in the acquisition of queue length and service rate information. Without any loss of generality, we take D_m's to be ordered in the following way:

$$0 \leq D_1 \leq D_2 \leq \ldots \leq D_M. \tag{2}$$

Hence,

$$r_{m,n} = \gamma_{m,n} \left(I_{n-D_m} \right), \quad n = 1, 2, \ldots; \; m = 1, 2, \ldots, M$$

where $\gamma_{m,n}$ are some measurable functions, with respect to which J will be minimized. For convenience, we introduce the new (appropriately shifted) variables

$$x_n := q_n - Q$$
$$u_{m,n} := r_{m,n} - a_m \mu \tag{3}$$

which will serve as the state and control, respectively. The queue dynamics (1) can be re-written in terms of these quantities as:

$$x_{n+1} = x_n + \sum_{m=1}^{M} u_{m,n} - \xi_n \tag{4}$$

$$\xi_{n+1} = \sum_{i=1}^{p} \alpha_i \xi_{n+1-i} + \phi_n \tag{5}$$

and the cost function J as:

$$J = \limsup_{N \to \infty} \frac{1}{N} E \left\{ \sum_{n=1}^{N} \left[(x_n)^2 + \sum_{m=1}^{M} c_m^2 (u_{m,n} - a_m \xi_n)^2 \right] \right\}.$$

What we have here is a decentralized optimal control problem with a partially nested information structure, and we know that the optimal solution to any finite horizon ($N < \infty$) version of this problem is linear in the available queue length and rate information [7]. If we ignore the delays in the system and solve the corresponding standard regulator problem, when it comes to incorporate the delays into the controllers again, we face the problem of non-uniqueness of the representation of the full-information (no delay) optimal controller, as discussed in [1]. Therefore, there are indeed many ways to define certainty-equivalent controllers and each such controller could lead to a different value of the performance index [1,4].

In the next section, we first derive the finite-horizon optimal controller for the two user case. We further show that the infinite horizon version of the two user problem admits a solution, which is linear in information variables. Section 4 discusses the extension of this derivation to M users. In particular, we present an easily implementable algorithm to calculate the optimal control action of each user for the finite horizon problem. Our simulation results strongly suggest that the infinite horizon version of the M user problem admits a solution, but we do not yet have a proof that the solution indeed exists in this case.

3   Derivation of Optimal Decentralized Flow Controllers for the Two User Case

3.1   Finite Horizon Optimal Controller

In this section, we consider the two user, N-stage problem where delay of user 1 is D_1 units, and that of user 2 is D_2 units. Define the relative delay of information between the users as

$$D := D_2 - D_1.$$

Here, without any loss of generality we can assume that $D_2 > D_1$, and we do not consider the trivial case $D_2 = D_1$, because in this case the problem can be reduced to a standard decentralized optimum control problem whose solution can be found easily. Our objective is to minimize the finite horizon cost

$$J^N = E \left\{ \sum_{n=D_2+1}^{N} \left[x_{n+1}^2 + c_1^2 (u_{1,n} - a_1 \xi_n)^2 + c_2^2 (u_{2,n} - a_2 \xi_n)^2 \right] \right.$$
$$\left. + \sum_{n=D_1+1}^{D_2} \left[x_{n+1}^2 + c_1^2 (u_{1,n} - \xi_n)^2 \right] \right\}$$

where $a_1 + a_2 = 1$ and c_1, c_2 are positive weights.

Derivation of the Controller We use a dynamic programming approach to obtain the solution to the above problem. More precisely, we start at time N and keep minimizing the expected cost backwards in time. In determining the control actions one can use completion of squares to see how the cost accumulates. Clearly, for $n = N, N-1, \ldots, N-D+1$ the minimization will be over a single variable, namely $u_{1,n}$. The reason for this is that the information field of user 1 at time n will be equivalent to that of user 2 at time $n+D$. Thus, until we reach the time $N-D$, there will be a sort of transient minimization process in which only the control actions of user 1 will be determined. At every step down to $N-D$, after we complete the squares in $u_{1,n}$ and pick the appropriate control, some terms remain which contribute to the cost of the succeeding step. If we study the problem carefully, it follows that these terms have a certain structure. To this end, we first minimize J with respect to $u_{1,N}$ at stage N. If we denote the cost to be minimized at stage n by J_n, we have, for $n = N$:

$$
J_N = x_{N+1}^2 + c_1^2 \left(u_{1,N} - a_1 \xi_N \right)^2
$$
$$
= \left(x_N + u_{1,N} + u_{2,N} - \xi_N \right)^2 + c_1^2 \left(u_{1,N} - a_1 \xi_N \right)^2 .
$$

Completion of squares yields

$$
J_N = \left(1 + c_1^2 \right) \left[u_{1,N} + \frac{1}{1 + c_1^2} \left(x_N + u_{2,N} \right) - \left(1 + \frac{c_1^2 \left(a_1 - 1 \right)}{1 + c_1^2} \right) \xi_N \right]^2
$$
$$
+ \frac{c_1^2}{1 + c_1^2} \left[u_{2,N} + x_N + \left(a_1 - 1 \right) \xi_N \right]^2 \tag{6}
$$

where $u_{1,N}$ must be selected such that the expected value of the first additive term in (6) is minimized. In fact, what we are interested in is the term that remains after the completion of squares, because, as we will see shortly, these terms determine the structure of the controllers when a steady state is reached. Let R_n denote these remaining terms at stage n. Thus the second additive term in (6) is R_N. In picking the control $u_{1,N}$, we do not need to consider the second term in J_N, since it will be taken care of at the next stage. Note that, user 1 at time N knows the control $u_{2,N}$, because $u_{1,N}$ has access to the information field of $u_{2,N}$. Therefore, the only difficulty in this minimization arises from the unknowns x_N and ξ_N. Since ϕ_n's are independent random variables, the best $u_{1,N}$ can do is to replace ξ_n's for $n = N, N-1, \ldots, N-D_1+1$ with their best estimates conditioned on its own information field. That is,

$$
\hat{\xi}_{1,n|N-D_1}^N = E \left\{ \xi_n \mid I_{N-D_1} \right\}, \quad n = N, N-1, \ldots, N-D_1+1. \tag{7}
$$

The reason for this can be seen if we substitute for x_N from the state equation (4). The result is

$$
J_N = (1 + c_1^2) \left[u_{1,N} - \left(1 + \frac{c_1^2 (a_1 - 1)}{1 + c_1^2} \right) \xi_N \right.
$$

$$
+ \frac{1}{1 + c_1^2} \left(x_{N-D_1} + u_{2,N} + \sum_{n=1}^{D_1} (u_{1,N-n} + u_{2,N-n} - \xi_{N-n}) \right) \right]^2 . \quad (8)
$$

In (8) the only unknowns to $u_{1,N}$ are $\xi_N, \xi_{N-1}, \ldots, \xi_{N-D_1+1}$. We already know that ξ_n's have a linear dynamics given by (5). Thus, (8) can be rewritten as

$$
J_N = (1 + c_1^2) \left[u_{1,N} - \frac{1}{1 + c_1^2} \sum_{n=1}^{D_1-1} \xi_{N-n} - \left(1 + \frac{c_1^2 (a_1 - 1)}{1 + c_1^2} \right) \xi_N \right.
$$

$$
+ \frac{1}{1 + c_1^2} \left(x_{N-D_1} + u_{2,N} + \xi_{N-D_1} + \sum_{n=1}^{D_1} (u_{1,N-n} + u_{2,N-n}) \right) \right]^2 .
$$

One can use (5) to express the ξ_n's in terms of ξ_{N-D_1} plus some zero mean random variable with a known variance. Thus, the best decision function $u_{1,N}$ is solely determined by the first term in (6) with unknown states replaced from the state equation (4) and unknown ξ_n's replaced with their best estimates given by (7). Our task at stage N ends with stating what R_N is

$$
R_N = \gamma_0 [u_{2,N} + \rho_{0,1} x_N + (a_1 - 1)\xi_N]^2
$$

where $\gamma_0 = \frac{c_1^2}{1 + c_1^2}$ and $\rho_{0,1} = 1$.

The next stage is $n = N - 1$. Assuming the relative delay of information between the users is larger than one time unit, we complete the squares in $u_{1,N-1}$ only. The cost functional to be minimized is

$$
J_{N-1} = x_N^2 + c_1^2 (u_{1,N-1} - a_1 \xi_N)^2 + R_N .
$$

Completion of squares results in three additive terms and two of them are transferred to the next stage. The first additive term determines the control $u_{1,N-1}$ at this stage. As in the previous step, the best $u_{1,N-1}$ can do is to replace ξ_n's for $n = N - 1, N - 2, \ldots, N - D_1$ with their best estimates conditioned on its information field. These estimates are given by

$$
\hat{\xi}_{1,n|N-D_1-1}^{N-1} = E \{ \xi_n \mid I_{N-D_1-1} \}, \quad n = N - 1, N - 2, \ldots, N - D_1,
$$

and R_{N-1} can be found to be

$$
R_{N-1} = \gamma_1 [u_{2,N} + \rho_{1,1} x_{N-1} + \rho_{1,1} u_{2,N-1} + \rho_{1,1} (a_1 - 1) \xi_{N-1} + (a_1 - 1)\xi_N]^2
$$
$$
+ \gamma_0 [u_{2,N-1} + \dot{\rho}_{0,1} x_{N-1} + (a_1 - 1)\xi_{N-1}]^2 .
$$

We proceed in this manner until we reach the time $N - D + 1$. From this point on, we start minimizing J_n's over two independent variables, namely $u_{1,n}$ and $u_{2,n+D}$. It is easy to calculate R_n's for the so-called transient part of this minimization process extending from time N to $N - D + 1$. For example, R_{N-2} is

$$
\begin{aligned}
R_{N-2} = \gamma_2 [& u_{2,N} + \rho_{2,1} x_{N-2} + \rho_{2,1} u_{2,N-2} + \rho_{2,1}(a_1 - 1)\xi_{N-2} \\
& + \rho_{2,2} u_{2,N-1} + \rho_{2,2}(a_1 - 1)\xi_{N-1} + (a_1 - 1)\xi_N]^2 \\
+ \gamma_1 [& u_{2,N-1} + \rho_{1,1} x_{N-2} + \rho_{1,1} u_{2,N-2} \\
& + \rho_{1,1}(a_1 - 1)\xi_{N-2} + (a_1 - 1)\xi_{N-1}]^2 \\
+ \gamma_0 [& u_{2,N-2} + \rho_{0,1} x_{N-2} + (a_1 - 1)\xi_{N-2}]^2
\end{aligned}
$$

and R_{N-3} equals

$$
\begin{aligned}
R_{N-3} = \gamma_3 [& u_{2,N} + \rho_{3,1} x_{N-3} + \rho_{3,1} u_{2,N-3} + \rho_{3,1}(a_1 - 1)\xi_{N-3} \\
& + \rho_{3,2} u_{2,N-2} + \rho_{3,2}(a_1 - 1)\xi_{N-2} + \rho_{3,3} u_{2,N-1} \\
& + \rho_{3,3}(a_1 - 1)\xi_{N-1} + (a_1 - 1)\xi_N]^2 \\
+ \gamma_2 [& u_{2,N-1} + \rho_{2,1} x_{N-3} + \rho_{2,1} u_{2,N-1} + \rho_{2,1}(a_1 - 1)\xi_{N-3} \\
& + \rho_{2,2} u_{2,N-2} + \rho_{2,2}(a_1 - 1)\xi_{N-2} + (a_1 - 1)\xi_{N-1}]^2 \\
+ \gamma_1 [& u_{2,N-2} + \rho_{1,1} x_{N-3} + \rho_{1,1} u_{2,N-3} \\
& + \rho_{1,1}(a_1 - 1)\xi_{N-3} + (a_1 - 1)\xi_{N-2}]^2 \\
+ \gamma_0 [& u_{2,N-3} + \rho_{0,1} x_{N-3} + (a_1 - 1)\xi_{N-3}]^2.
\end{aligned}
$$

Note that the evolution of R_n obeys a certain structure. One can exploit this structure to determine the constants γ_d's and ρ_d's:

$$
\tau_{-1} = 1 + c_1^2, \quad \tau_d = 1 + c_1^2 + \sum_{n=0, D \geq 2}^{d} \gamma_n \rho_{n,1}^2, \gamma_{-1} = 0, \quad \gamma_0 = \frac{c_1^2}{1 + c_1^2},
$$

$$
\rho_{0,1} = 1, \quad \gamma_{d+1} = \gamma_d - \frac{\gamma_d^2 \rho_{d,1}^2}{\tau_d}, \quad d = 0, 1, \ldots, D - 2, \; D \geq 2,
$$

$$
\rho_{d+1,1} = \frac{\gamma_d \rho_{d,1}}{\gamma_{d+1}} \left(1 - \frac{1 + \gamma_0 \rho_{0,1} + \gamma_1 \rho_{1,1}^2 + \cdots + \gamma_d \rho_{d,1}^2}{\tau_d} \right),
$$
$$
d = 0, 1, \ldots, D - 2, \; D \geq 2,
$$

$$
\rho_{d+1,2} = \frac{\gamma_d \rho_{d,1}}{\gamma_{d+1}} \left(1 - \frac{\gamma_0 \rho_{0,1} + \gamma_1 \rho_{1,1}^2 + \cdots + \gamma_d \rho_{d,1}^2}{\tau_d} \right),
$$
$$
d = 1, 2, \ldots, D - 2, \; D \geq 3,
$$

$$
\rho_{d+1,3} = \frac{\gamma_d \rho_{d,2}}{\gamma_{d+1}} - \frac{\gamma_d \rho_{d,1}}{\gamma_{d+1}} \left(\frac{\gamma_1 \rho_{1,1} + \gamma_2 \rho_{2,1} \rho_{2,2} + \cdots + \gamma_d \rho_{d,1} \rho_{d,2}}{\tau_d} \right),
$$
$$
d = 2, 3, \ldots, D - 2, \; D \geq 4,
$$

$$\vdots$$

$$\rho_{d+1,k+1} = \frac{\gamma_d \rho_{d,k}}{\gamma_{d+1}} - \frac{\gamma_d \rho_{d,1}}{\gamma_{d+1}} \left(\frac{\gamma_{k-1}\rho_{k-1,1} + \gamma_k \rho_{k,1}\rho_{k,k} + \cdots + \gamma_d \rho_{d,1}\rho_{d,k}}{\tau_d} \right),$$
$$d = k, k+1, \ldots, D-2, \ D \geq k+2, \ k = 4, \ldots, D-2.$$

For $n = N, N-1, \ldots, N-D+1$, the cost functionals we minimize have the form

$$J_n = x_{n+1}^2 + c_1^2 (u_{1,n} - a_1 \xi_n)^2 + R_{n+1} \tag{9}$$

with $R_{N+1} = 0$. The optimal control for this time interval can be written as

$$u_{1,N-d}^* = -\sigma_{0,d} \left(\hat{x}_{1,N-d|N-D_1-d}^{N-d} + u_{2,N-d} \right)$$

$$- \sum_{k=1, d \neq 0}^{d} \sigma_{k,d} \left(u_{2,N-d+k} + (a_1 - 1) \hat{\xi}_{1,N-d+k|N-D_1-d}^{N-d} \right)$$

$$+ (a_1 - \sigma_{0,d}(a_1 - 1)) \hat{\xi}_{1,N-d|N-D_1-d}^{N-d}, \ d = 0, \ldots, D-1, \tag{10}$$

where

$$\sigma_{0,d} = \frac{\tau_{d-1} - c_1^2}{\tau_{d-1}}, \ d \geq 0$$

$$\sigma_{1,d} = \frac{\gamma_0 \rho_{0,1} + \gamma_1 \rho_{1,1}^2 + \cdots + \gamma_{d-1}\rho_{d-1,1}^2}{\tau_{d-1}}, \ d \geq 1,$$

$$\sigma_{2,d} = \frac{\gamma_1 \rho_{1,1} + \gamma_2 \rho_{2,1}\rho_{2,2} + \cdots + \gamma_{d-1}\rho_{d-1,1}\rho_{d-1,2}}{\tau_{d-1}}, \ d \geq 2,$$

$$\vdots$$

$$\sigma_{k,d} = \frac{\gamma_{k-1}\rho_{k-1,1} + \gamma_k \rho_{k,1}\rho_{k,k} + \cdots + \gamma_{d-1}\rho_{d-1,1}\rho_{d-1,k}}{\tau_{d-1}}, \ d \geq k,$$

$$k = 4, \ldots, D-1.$$

In the above set of equations, $\hat{\xi}_{1,n|k}^l$ denotes source 1's estimate at time l of the value of ξ at time n based on I_k. A similar interpretation holds for $\hat{x}_{1,n|k}^l$.

Now, the next step is to calculate the optimal control laws for $D_2 + 1 \leq n \leq N - D$. First, we observe that the cost functional to be minimized at any stage after D steps back in time is

$$J_n = [x_n + u_{1,n} + u_{2,n} - \xi_n]^2 + c_1^2 [u_{1,n} - a_1 \xi_n]^2 + c_2^2 [u_{2,n+D} - a_2 \xi_{n+D}]^2$$

$$+ \eta_n [u_{2,n+D} + \rho_{D-1,1}x_{n+1} + \rho_{D-1,1}u_{2,n+1} + \rho_{D-1,1}(a_1 - 1)\xi_{n+1}$$

$$+ \rho_{D-1,2}u_{2,n+2} + \rho_{D-1,2}(a_1 - 1)\xi_{n+2}$$

$$+ \cdots + \rho_{D-1,D-1}u_{2,n+D-1} + \rho_{D-1,D-1}(a_1 - 1)\xi_{n+D-1}$$

$$+ (a_1 - 1)\xi_{n+D}]^2$$

$$+ \gamma_{D-2}[u_{2,n+D-1} + \rho_{D-2,1}x_{n+1} + \rho_{D-2,1}u_{2,n+1}$$
$$+ \rho_{D-2,1}(a_1 - 1)\xi_{n+1} + \rho_{D-2,2}u_{2,n+2}$$
$$+ \rho_{D-2,2}(a_1 - 1)\xi_{n+2} + \cdots + \rho_{D-2,D-2}u_{2,n+D-2}$$
$$+ \rho_{D-2,D-2}(a_1 - 1)\xi_{2,n+D-2} + (a_1 - 1)\xi_{n+D-1}]^2$$
$$+ \cdots + \gamma_1[u_{2,n+2} + \rho_{1,1}x_{n+1} + \rho_{1,1}u_{2,n+1} + \rho_{1,1}(a_1 - 1)\xi_{n+1}$$
$$+ (a_1 - 1)\xi_{n+2}]^2$$
$$+ \gamma_0[u_{2,n+1} + \rho_{0,1}x_{n+1} + (a_1 - 1)\xi_{n+1}]^2 \tag{11}$$

where

$$\eta_{N-D} = \gamma_{D-1}. \tag{12}$$

A recursive formula to calculate η_n's can be found after some algebraic manipulations:

$$\eta_{n-1} = \frac{A_1 \eta_n + A_0}{B_1 \eta_n + B_0}, \quad n = N - D, N - D - 1, \ldots, D_2 + 1 \tag{13}$$

where

$$A_1 = (\gamma_{D-2} + c_2^2 \rho_{D-1,D-1}^2)\left(1 + c_1^2 + \sum_{k=3,D\geq3}^{D} \gamma_{D-k}\rho_{D-k,1}^2\right)$$
$$\times c_2^2 \gamma_{D-2}(\rho_{D-1,1} - \rho_{D-2,1}\rho_{D-1,D-1})^2$$

$$A_0 = c_2^2 \gamma_{D-2}\left(1 + c_1^2 + \sum_{k=3,D\geq3}^{D} \gamma_{D-k}\rho_{D-k,1}^2\right)$$

$$B_1 = \left(1 + c_1^2 + c_2^2 \rho_{D-1,1}^2 + \sum_{k=2,D\geq2}^{D} \gamma_{D-k}\rho_{D-k,1}^2\right)$$

$$B_0 = c_2^2 \left(1 + c_1^2 + \sum_{k=2,D\geq2}^{D} \gamma_{D-k}\rho_{D-k,1}^2\right). \tag{14}$$

Note that all these four quantities are positive. Now, define $\psi_{n|D}$ for a given relative delay D as

$$\psi_{n|D} := 1 + c_1^2 + \eta_n \rho_{D-1,1}^2 + \gamma_{D-2}\rho_{D-2,1}^2 + \cdots + \gamma_0 \rho_{0,1}^2.$$

The optimal controls can be found by minimizing (11) over $u_{1,n}$ and $u_{2,n+D}$. Before writing down the complete solution, let us introduce some definitions.

For given n and D, let $\pi_{k|n,D}$ be defined by

$$\pi_{0|n,D} := \frac{\eta_n \rho_{D-1,1}}{\psi_{n|D}}$$

$$\pi_{1|n,D} := \frac{\eta_n \rho_{D-1,1}\rho_{D-1,D-1} + \gamma_{D-2}\rho_{D-2,1}}{\psi_{n|D}}$$

$$\pi_{k|n,D} := \frac{\eta_n \rho_{D-1,1}\rho_{D-1,D-k} + \sum_{j=2}^{k}\left[\gamma_{D-j}\rho_{D-j,1}\rho_{D-j,D-k}\right]}{\psi_{n|D}}$$

$$+ \frac{\gamma_{D-k-1}\rho_{D-k-1,1}}{\psi_{n|D}}, \quad k = 2, 3, \ldots, D-1,$$

$$\pi_{D|n,D} := \pi_{D-1|n,D} + \frac{1}{\psi_{n|D}}$$

and let $\lambda_{n|D}$ be defined as

$$\lambda_{n|D} := c_2^2 + \eta_n - \frac{\eta_n^2 \rho_{D-1,1}^2}{\psi_{n|D}}.$$

Then, the optimum control $u_{2,n+D}^*$ becomes

$u_{2,n+D}^*$

$$= -\frac{\eta_n \rho_{D-1,1} - \psi_{n|D}\pi_{0|n,D}\pi_{D|n,D}}{\lambda_{n|D}} \left(\hat{x}_{2,n|n-D_1}^{n+D} + u_{2,n} + (a_1 - 1)\,\hat{\xi}_{2,n|n-D_1}^{n+D}\right)$$

$$- (a_1 - 1)\,\hat{\xi}_{2,n+D|n-D_1}^{n+D}$$

$$- \sum_{k=1,D\geq 1}^{D-1} \left[\frac{\eta_n \rho_{D-1,k} - \psi_{n|D}\pi_{0|n,D}\pi_{D-k|n,D}}{\lambda_{n|D}}\right.$$

$$\left.\times \left(u_{2,n+k} + (a_1 - 1)\,\hat{\xi}_{2,n+k|n-D_1}^{n+D}\right)\right], \quad n = N - D, \ldots, D_2 + 1. \quad (15)$$

Similarly, one can write down the control $u_{1,n}^*$ as

$$u_{1,n}^* = -\pi_{0|n,D}\left(u_{2,n+D}^* + (a_1 - 1)\hat{\xi}_{1,n+D|n-D_1}^{n}\right)$$

$$- \sum_{k=1}^{D-1}\left[\pi_{k|n,D}\left(u_{2,n+D-k} + (a_1 - 1)\hat{\xi}_{1,n+D-k|n-D_1}^{n}\right)\right]$$

$$- \pi_{D|n,D}\left(\hat{x}_{1,n|n-D_1}^{n} + u_{2,n}\right)$$

$$+ \left(1 + \frac{c_1^2(a_1 - 1)}{\psi_{n|D}}\right)\hat{\xi}_{1,n|n-D_1}^{n}, \quad n = N - D, \ldots, D_2 + 1. \quad (16)$$

Note that, $u_{2,n+D}^*$ appears in the expression for $u_{1,n}^*$, but this is not necessary, as one can substitute (15) into (16) and have a formula for $u_{1,n}^*$ expressed only in terms of its own information variables. The calculation of the controllers for the last D steps of the problem is omitted, but it can be carried out without much effort.

Certainty Equivalence The optimal controller derived above is certainty-equivalent in the sense that if users 1 and 2 had access to perfect state information, they would simply replace the estimated variables in (15) and (16) with their actual values. In fact, this follows from the construction of the controller. In the derivations, at each stage we replaced the unknown states with their best estimates conditioned on users' own information fields at that stage. Hence, if we assume that these unknown states are known to the users, the best they can do is to use this new information in their expressions of control. As we mentioned earlier, there are many ways in which a certainty-equivalent controller can be defined and each such controller leads to a different value for the cost [1], [4]. Of course, among these different controllers (each one being a particular *representation* of the perfect state information controller), the one we constructed above has the lowest possible cost due to its optimality.

3.2   Infinite Horizon Optimal Controller

We now return to the original infinite-horizon problem with two users. Our objective was actually to minimize the infinite-horizon cost

$$
J = \limsup_{N \to \infty} \frac{1}{N} E \left\{ \sum_{n=D_2+1}^{N} \left[x_{n+1}^2 + c_1^2 (u_{1,n} - a_1 \xi_n)^2 + c_2^2 (u_{2,n} - a_2 \xi_n)^2 \right] \right.
$$

$$
\left. + \sum_{n=D_1+1}^{D_2} \left[x_{n+1}^2 + c_1^2 (u_{1,n} - \xi_n)^2 \right] \right\}. \quad (17)
$$

Here we prefer to write lim sup instead of just lim, because we do not know yet whether the limit exists. To be able to prove the existence of the infinite-horizon optimal controller, we first show that the controller dynamics given by (12) and (13) converge as N tends to ∞. Next, we use this result to show that the optimal finite-horizon controller is stabilizing as $N \to \infty$. And finally, we prove that the optimum value of (17), denoted by J^*, equals to the limit of the optimum finite horizon average cost. Thus, the linear stationary policies which are given as the limit of the optimal policies (15) and (16) are optimal for the infinite-horizon problem.

Convergence of Controller Dynamics In an infinite horizon problem, one usually wants to know whether the recursively defined sequence (13) has a limit point for arbitrary values of system parameters. And if so, we have to make sure that this limit point is unique. In what follows we address these questions. First of all, note that, A_1, A_0, B_1 and B_0 are all positive constants. The initial condition of (13) being a positive constant, η_n's remain positive for all n. To investigate the existence and uniqueness of the limit we formulate the problem as follows. The sequence of points given by (13) is computed in reverse time by a formula of the form

$$
\eta_{n-1} = F(\eta_n), \quad n = N - D, N - D - 1, \dots, D_2 + 1
$$

where

$$F(\eta) = \frac{A_1 \eta + A_0}{B_1 \eta + B_0}. \tag{18}$$

Note that, F is positive and continuous for $\eta \geq 0$, since A_1, A_0, B_1, B_0 are all positive. For convenience, as $N \to \infty$, we take the sequence $\{\eta_n\}$ to be defined in forward time as

$$\eta_{n+1} = F(\eta_n) \tag{19}$$

with the initial value $\eta_0 = \gamma_{D-1}$. This new definition does not change anything about the existence and uniqueness of the limit. Now, assume that the sequence $\{\eta_n\}$ has at least one limit point denoted by η_∞. If so, we must have η_n's converging to this number, and hence to a solution of the equation:

$$\eta_\infty = \frac{A_1 \eta_\infty + A_0}{B_1 \eta_\infty + B_0}$$

which is equivalent to the following quadratic equation in η_∞:

$$B_1 \eta_\infty^2 + (B_0 - A_1) \eta_\infty - A_0 = 0 \tag{20}$$

whose discriminant is

$$\Delta = (B_0 - A_1)^2 + 4B_1 A_0 > 0. \tag{21}$$

Thus, we have two real roots. The signs of the roots are opposite, because their product, given by $-\frac{A_0}{B_1}$, is negative. As a result, we can conclude that we have a single positive real root of (20), designated by η_∞^+. Note that, η_∞^+ is a *fixed point* of (18), and is unique on the interval $[0, \infty)$. By the *contraction mapping theorem* [8, pp. 80–82], this point is indeed the unique limit of every sequence obtained from (19) with any nonnegative starting point if F is a contraction mapping on $[0, \infty) \cup \{\infty\}$. We proceed by showing that F is indeed a contraction on this closed interval. To this end we seek a number $\beta < 1$, such that

$$|F(x) - F(y)| \leq \beta |x - y|, \quad \forall x, y \geq 0. \tag{22}$$

Plugging (18) into (22) yields

$$\frac{|x - y| |A_0 B_1 - A_1 B_0|}{(B_1 x + B_0)(B_1 y + B_0)} \leq \beta |x - y|.$$

Cancelling $|x - y|$'s and using the facts that the minima of $(B_1 x + B_0)$ and $(B_1 y + B_0)$ over $[0, \infty)$ occur at $x = 0$ and $y = 0$ respectively, and B_0^2 is positive, we can rewrite the condition of contraction as

$$|A_0 B_1 - A_1 B_0| - B_0^2 < 0.$$

Next, we substitute for A_0, B_0, A_1 and B_1 from (14) and after some algebraic manipulations we get

$$
\left[(\rho_{D-1,D-1} - 1) \left(\gamma_{D-2}\rho_{D-2,1}^2 + 1 + c_1^2 + \sum_{k=3,D\geq 3}^{D} \gamma_{D-k}\rho_{D-k,1}^2 \right) \right.
$$

$$
\left. - \gamma_{D-2}\rho_{D-1,1}\rho_{D-2,1} \right]
$$

$$
\times \left[\rho_{D-1,D-1} \left(\gamma_{D-2}\rho_{D-2,1}^2 + 1 + c_1^2 + \sum_{k=3,D\geq 3}^{D} \gamma_{D-k}\rho_{D-k,1}^2 \right) \right.
$$

$$
+ \left(\gamma_{D-2}\rho_{D-2,1}^2 + 1 + c_1^2 + \sum_{k=3,D\geq 3}^{D} \gamma_{D-k}\rho_{D-k,1}^2 \right)
$$

$$
\left. - \gamma_{D-2}\rho_{D-1,1}\rho_{D-2,1} \right] < 0. \quad (23)
$$

The first multiplicative term in (23) is negative, because $(\rho_{D-1,D-1} - 1) < 0$ for any D, and the second multiplicative term is positive, since we have $\gamma_{D-2}\rho_{D-1,1}\rho_{D-2,1} < 1$ by construction. As a result, the inequality in (23) is satisfied for arbitrary values of system parameters, but (23) is equivalent to (22), and thus F is a contraction mapping. Hence, the contraction mapping theorem applies and the fixed point η_∞^+ turns out to be the unique limit point of the sequence $\{\eta_n\}$. The value of η_∞^+ can be calculated as the positive root of (20):

$$
\eta_\infty^+ = \frac{(A_1 - B_0) + \sqrt{(B_0 - A_1)^2 + 4B_1 A_0}}{2B_1}.
$$

Now, the stationary optimal policies can be found by plugging the limiting value of the sequence $\{\eta_n\}$ into (15) and (16). We will see shortly that these stationary policies indeed constitute the solution of the infinite-horizon version of the problem, but first we prove another useful result which is also used in showing the existence of a solution to the infinite-horizon problem.

Stabilizing Property of the Optimal Controller In this subsection, we investigate the stabilizing property of the optimal controller and show that as N goes to infinity, the average cost remains bounded. An immediate consequence of this result is that neither the shifted queue dynamics nor the user rates can blow up when the optimal controller is applied. In [1] it was shown that the average cost is bounded for two suboptimal certainty-equivalent controllers, referred to as *Controller 1* and *Controller 2*. For instance, for Controller 1 we have

$$J^* \le \lim_{N \to \infty} \frac{J^N}{N} \le C_1$$

where C_1 is the scalar given in Section 5 of [1]. The controller here being optimal leads to an average cost that is no larger. Therefore,

$$J^* \le \lim_{N \to \infty} \frac{J^N}{N} \le C_{opt} \le C_1$$

which proves the stabilizability of the optimal controller.

Here we present a more direct proof of this fact by calculating the exact value of the cost per stage. Starting at time N, until we reach the time $N - D$, the control is given by (10), and for any given D the cost of these first D stages, denoted by L_∞, can be found by plugging (10) into (9) and it will depend on the parameters of the AR process as well as the control (10). It is clear that L_∞ is finite, because we have only a finite number of steps until the time $N - D + 1$. We also note that the optimal control laws for $D_2 + 1 \le n \le N - D$ were found by minimizing J_n's over $u_{1,n}$ and $u_{2,n+D}$. If we complete J_n to squares in $u_{1,n}$ and $u_{2,n+D}$ and substitute for the optimal controls (15) and (16), the expected value of the resulting expression gives us a cost that we cannot avoid due to the delays in the system, plus some remaining terms that are transferred to the next stage of the minimization process. These remaining terms are taken care of in minimizing J_{n-1}. Thus, the cost of stage n essentially equals to the expected value of the following quantity

$$\Omega_n = \psi_{n|D} \left[u_{1,n}^* + \sum_{k=1}^{D-1} \left[\pi_{k|n,D} (u_{2,n+D-k} + (a_1 - 1)\xi_{n+D-k}) \right] \right.$$

$$+ \pi_{0|n,D} \left(u_{2,n+D}^* + (a_1 - 1)\xi_{n+D} \right) + \pi_{D|n,D} (x_n + u_{2,n})$$

$$\left. - \left(1 + \frac{c_1^2 (a_1 - 1)}{\psi_{n|D}} \right) \xi_n \right]^2$$

$$+ \lambda_{n|D} \left[u_{2,n+D}^* + \frac{\eta_n \rho_{D-1,1} - \psi_{n|D} \pi_{0|n,D} \pi_{D|n,D}}{\lambda_{n|D}} (x_n + u_{2,n} + (a_1 - 1)\xi_n) \right.$$

$$+ (a_1 - 1)\xi_{n+D}$$

$$+ \sum_{k=1,D \ge 1}^{D-1} \left[\frac{\eta_n \rho_{D-1,k} - \psi_{n|D} \pi_{0|n,D} \pi_{D-k|n,D}}{\lambda_{n|D}} \right.$$

$$\left. \left. \times (u_{2,n+k} + (a_1 - 1)\xi_{n+k}) \right] \right]^2 .$$

If we substitute for the optimal controls $u^*_{1,n}$ and $u^*_{2,n+D}$, we get

$$
\Omega_n = \psi_{n|D} \Bigg[\pi_{0|n,D} \left(a_1 - 1\right) \left(\xi_{n+D} - \hat{\xi}^n_{1,n+D|n-D_1}\right)
$$

$$
+ \pi_{D|n,D} \left(x_n - \hat{x}^n_{1,n|n-D-1}\right)
$$

$$
+ \sum_{k=1}^{D-1} \pi_{k|n,D} \left(a_1 - 1\right) \left(\xi_{n+D-k} - \hat{\xi}^n_{1,n+D-k|n-D_1}\right)
$$

$$
- \left(1 + \frac{c_1^2 \left(a_1 - 1\right)}{\psi_{n|D}}\right) \left(\xi_n - \hat{\xi}^n_{1,n|n-D-1}\right) \Bigg]^2
$$

$$
+ \lambda_{n|D} \Bigg[\frac{\eta_n \rho_{D-1,1} - \psi_{n|D} \pi_{0|n,D} \pi_{D|n,D}}{\lambda_{n|D}}
$$

$$
\times \left(\left(x_n - \hat{x}^{n+D}_{2,n|n-D_1}\right) + \left(a_1 - 1\right) \left(\xi_n - \hat{\xi}^{n+D}_{2,n|n-D_1}\right) \right)
$$

$$
+ \left(a_1 - 1\right) \left(\xi_{n+D} - \hat{\xi}^{n+D}_{2,n+D|n-D_1}\right)
$$

$$
+ \sum_{k=1,D\geq 1}^{D-1} \frac{\eta_n \rho_{D-1,k} - \psi_{n|D} \pi_{0|n,D} \pi_{D-k|n,D}}{\lambda_{n|D}}
$$

$$
\times \left(a_1 - 1\right) \left(\xi_{n+k} - \hat{\xi}^{n+D}_{2,n+k|n-D_1}\right) \Bigg]^2 .
$$

Hence, to calculate the cost at stage n, first we need to express the terms involving the difference between estimated variables and their actual values in a convenient form. Following along the lines of Section 5 of [1], one can show that the expected value of Ω_n can equivalently be written as

$$
E\{\Omega_n\} = \varphi_n E\{\phi^2\} = \varphi_n \sigma_\phi^2 \tag{24}
$$

where φ_n is the sequence obtained from Ω_n by using the fact that the estimation errors are linear in $\{\phi_n\}$'s which are $i.i.d$ with variance σ_ϕ^2. For a more detailed reasoning behind this idea see [1], Section 5. Now, as n tends to $D_2 + 1$, while N tending to ∞, we know that η_n's converge to the unique number η_∞^+. To investigate the limiting behavior of φ_n, we need to consider that of $\pi_{l|n,D}, l = 0, \ldots, D$, $\psi_{n|D}$ and $\lambda_{n|D}$, because φ_n for each n is just a continuous function of these sequences. Hence, if one can show that these three sequences converge, then by continuity φ_n's will converge as well. To this end, we first recall that

$$
\psi_{n|D} = 1 + c_1^2 + \eta_n \rho_{D-1,1}^2 + \gamma_{D-2} \rho_{D-2,1}^2 + \ldots + \gamma_0 \rho_{0,1}^2, \tag{25}
$$

$$\pi_{0|n,D} := \frac{\eta_n \rho_{D-1,1}}{\psi_{n|D}}$$

$$\pi_{1|n,D} := \frac{\eta_n \rho_{D-1,1}\rho_{D-1,D-1} + \gamma_{D-2}\rho_{D-2,1}}{\psi_{n|D}}$$

$$\pi_{k|n,D} := \frac{\eta_n \rho_{D-1,1}\rho_{D-1,D-k} + \sum_{j=2}^{k}[\gamma_{D-j}\rho_{D-j,1}\rho_{D-j,D-k}]}{\psi_{n|D}}$$

$$+ \frac{\gamma_{D-k-1}\rho_{D-k-1,1}}{\psi_{n|D}}, \quad k = 2,3,\dots,D-1,$$

$$\pi_{D|n,D} := \pi_{D-1|n,D} + \frac{1}{\psi_{n|D}} \tag{26}$$

and

$$\lambda_{n|D} = c_2^2 + \eta_n - \frac{\eta_n^2 \rho_{D-1,1}^2}{\psi_{n|D}}. \tag{27}$$

Taking the limit of both sides of (25), (26) and (27) we see that

$$\psi_{\infty|D} = 1 + c_1^2 + \eta_{\infty}^{+}\rho_{D-1,1}^2 + \gamma_{D-2}\rho_{D-2,1}^2 + \cdots + \gamma_0 \rho_{0,1}^2,$$

$$\pi_{0|\infty,D} := \frac{\eta_{\infty}^{+}\rho_{D-1,1}}{\psi_{\infty|D}}$$

$$\pi_{1|\infty,D} := \frac{\eta_{\infty}^{+}\rho_{D-1,1}\rho_{D-1,D-1} + \gamma_{D-2}\rho_{D-2,1}}{\psi_{\infty|D}}$$

$$\pi_{k|\infty,D} := \frac{\eta_{\infty}^{+}\rho_{D-1,1}\rho_{D-1,D-k} + \sum_{j=2}^{k}[\gamma_{D-j}\rho_{D-j,1}\rho_{D-j,D-k}]}{\psi_{\infty|D}}$$

$$+ \frac{\gamma_{D-k-1}\rho_{D-k-1,1}}{\psi_{\infty|D}}, \quad k = 2,3,\dots,D-1,$$

$$\pi_{D|\infty,D} := \pi_{D-1|\infty,D} + \frac{1}{\psi_{\infty|D}}$$

and

$$\lambda_{\infty|D} = c_2^2 + \eta_{\infty}^{+} - \frac{\eta_{\infty}^{+2}\rho_{D-1,1}^2}{\psi_{\infty|D}}.$$

Since all three limits exist, the sequence $\{\varphi_n\}$ converges to a real number say, φ_{∞}. Using this result we can write the cost J^N as

$$J^N = \sum_{n=D_1+1}^{N-D} E\{\Omega_n\} + L_{\infty} = L_{\infty} + \sigma_{\phi}^2 \sum_{n=D_1+1}^{N-D} \varphi_n.$$

Hence, the average cost as $N \to \infty$ is equal to

$$\lim_{N\to\infty} \frac{J^N}{N} = \lim_{N\to\infty} \frac{1}{N}\left(L_{\infty} + \sigma_{\phi}^2 \sum_{n=D_1+1}^{N-D} \varphi_n\right).$$

We can actually evaluate this limit by using the following fact.

Fact: Let $\{a_n\}$ be a convergent sequence with limit a, and let $m < \infty$ be an arbitrary scalar. Then, the following infinite sum

$$\lim_{N \to \infty} \frac{1}{N} \sum_{n=m}^{N} a_n$$

converges to the real number a.

Therefore, we must have

$$\lim_{N \to \infty} \frac{J^N}{N} = \sigma_\phi^2 \varphi_\infty$$

which proves that the average cost is bounded if the limit of the finite-horizon optimal controller is applied to the system. This does not necessarily mean that this limiting stationary controller minimize the average infinite horizon cost. In the next subsection, however, we will show that this is indeed the case.

Infinite Horizon Controller Having shown that the controller dynamics converge and the limiting case of the finite-horizon optimal controller is stabilizing, in this section we prove a stronger result, which enables us to calculate the infinite-horizon optimal controller as the limit of the finite-horizon one. Let M be a positive integer and let

$$\Pi = \{\Gamma_{D_1+1}, \Gamma_{D_1+2}, \dots, \Gamma_M, \Gamma_{M+1}, \dots\}$$

be any admissible policy where each Γ_m consists of two elements: $u_{1,m}$ and $u_{2,m+D}$. We shall denote the optimal policy (15)-(16) by

$$\Pi^* = \{\Gamma_{D_1+1}^*, \Gamma_{D_1+2}^*, \dots, \Gamma_M^*, \Gamma_{M+1}^*, \dots\}.$$

Without any loss of generality, we assume that M is such that the transient part of the dynamic programming is over. First, recall from Section 2 that

$$\min_{u_{1,M}, u_{2,M+D}} E\{S_M + H_M\} = E\{\Omega_M\} + H_{M-1} \tag{28}$$

where

$$S_n := x_{n+1}^2 + c_1^2 [u_{1,n} - a_1 \xi_n]^2 + c_2^2 [u_{2,n+D} - a_2 \xi_{n+D}]^2$$

and

$$
\begin{aligned}
H_n := \eta_n[&u_{2,n+D} + \rho_{D-1,1}x_{n+1} + \rho_{D-1,1}u_{2,n+1} + \rho_{D-1,1}(a_1 - 1)\xi_{n+1} \\
&+ \rho_{D-1,2}u_{2,n+2} + \rho_{D-1,2}(a_1 - 1)\xi_{n+2} \\
&+ \cdots + \rho_{D-1,D-1}u_{2,n+D-1} + \rho_{D-1,D-1}(a_1 - 1)\xi_{n+D-1} \\
&+ (a_1 - 1)\xi_{n+D}]^2 \\
+ \gamma_{D-2}[&u_{2,n+D-1} + \rho_{D-2,1}x_{n+1} + \rho_{D-2,1}u_{2,n+1} \\
&+ \rho_{D-2,1}(a_1 - 1)\xi_{n+1} + \rho_{D-2,2}u_{2,n+2} \\
&+ \rho_{D-2,2}(a_1 - 1)\xi_{n+2} + \ldots + \rho_{D-2,D-2}u_{2,n+D-2} \\
&+ \rho_{D-2,D-2}(a_1 - 1)\xi_{2,n+D-2} + (a_1 - 1)\xi_{n+D-1}]^2 \\
+ \cdots + \gamma_1[&u_{2,n+2} + \rho_{1,1}x_{n+1} + \rho_{1,1}u_{2,n+1} + \rho_{1,1}(a_1 - 1)\xi_{n+1} \\
&+ (a_1 - 1)\xi_{n+2}]^2 \\
+ \gamma_0[&u_{2,n+1} + \rho_{0,1}x_{n+1} + (a_1 - 1)\xi_{n+1}]^2.
\end{aligned}
$$

Using (24), we can rewrite (28) as

$$
\min_{u_{1,M}, u_{2,M+D}} E\{S_M + H_M\} = \sigma_\phi^2 \varphi_M + H_{M-1}. \tag{29}
$$

At this point, for convenience, we introduce a mapping $T_{\Gamma_n}(\cdot)$ which simply evaluates the expected value of its argument when the control policy is given by Γ_n. We have, from (29)

$$
T_{\Gamma_M}(S_M + H_M) \geq \sigma_\phi^2 \varphi_M + h_{M-1}. \tag{30}
$$

By applying $T_{\Gamma_{M-1}}$ to both sides of (30) we see that

$$
\begin{aligned}
T_{\Gamma_{M-1}}(T_{\Gamma_M}(S_M + H_M) + S_{M-1}) &\geq T_{\Gamma_{M-1}}\left(\sigma_\phi^2 \varphi_M + H_{M-1} + S_{M-1}\right) \\
&= \sigma_\phi^2 \varphi_M + T_{\Gamma_{M-1}}(H_{M-1} + S_{M-1}) \\
&\geq \sigma_\phi^2(\varphi_M + \varphi_{M-1}) + H_{M-2}
\end{aligned}
$$

where the first inequality follows from a simple principle of optimality argument. Proceeding in the same manner, we finally obtain

$$
\begin{aligned}
T_{\Gamma_{D_1+1}}\left(T_{\Gamma_{D_1+2}}(\cdots(T_{\Gamma_M}(S_M + H_M) + \cdots) + S_{D_1+2}) + S_{D_1+1}\right) \\
\geq \sigma_\phi^2 \sum_{n=D_1+1}^{M} \varphi_n \tag{31}
\end{aligned}
$$

with equality if each Γ_n, $n = D_1 + 1, D_1 + 2, \ldots, M$ equals Γ_n^*. Now, the left-hand side of (31) is equal to the M-stage cost corresponding to the policy

$\Pi = \{\Gamma_{D_1+1}, \Gamma_{D_1+2}, \ldots, \Gamma_M\}$. In other words,

$$T_{\Gamma_{D_1+1}} \left(T_{\Gamma_{D_1+2}} \left(\cdots \left(T_{\Gamma_M} \left(S_M + H_M\right) + \cdots\right) + S_{D_1+2}\right) + S_{D_1+1}\right)$$

$$= E\left\{\sum_{n=D_2+1}^{M} \left[x_{n+1}^2 + c_1^2 \left(u_{1,n} - a_1\xi_n\right)^2 + c_2^2 \left(u_{2,n} - a_2\xi_n\right)^2\right]\right.$$

$$\left.+ \sum_{n=D_1+1}^{D_2} \left[x_{n+1}^2 + c_1^2 \left(u_{1,n} - \xi_n\right)^2\right] \middle| \Pi \right\}.$$

One can use this relation in (31) to get

$$E\left\{\sum_{n=D_2+1}^{M} \left[x_{n+1}^2 + c_1^2 \left(u_{1,n} - a_1\xi_n\right)^2 + c_2^2 \left(u_{2,n} - a_2\xi_n\right)^2\right]\right.$$

$$\left.+ \sum_{n=D_1+1}^{D_2} \left[x_{n+1}^2 + c_1^2 \left(u_{1,n} - \xi_n\right)^2\right] \middle| \Pi \right\}$$

$$\geq \sigma_\phi^2 \sum_{n=D_1+1}^{M} \varphi_n. \quad (32)$$

If we divide both sides of (32) by M and take the limit as M goes to ∞, we obtain the inequality

$$\lim_{M\to\infty} \frac{1}{M} E\left\{\sum_{n=D_2+1}^{M} \left[x_{n+1}^2 + c_1^2 \left(u_{1,n} - a_1\xi_n\right)^2 + c_2^2 \left(u_{2,n} - a_2\xi_n\right)^2\right]\right.$$

$$\left.+ \sum_{n=D_1+1}^{D_2} \left[x_{n+1}^2 + c_1^2 \left(u_{1,n} - \xi_n\right)^2\right] \middle| \Pi \right\}$$

$$\geq \lim_{M\to\infty} \frac{1}{M} \sigma_\phi^2 \sum_{n=D_1+1}^{M} \varphi_n = \sigma_\phi^2 \varphi_\infty$$

with equality if $\Pi = \Pi^*$. Thus, the infinite horizon cost J is lower bounded by the real number $\sigma_\phi^2 \varphi_\infty$ and this bound can be achieved if the policy is that of the limit of the optimal finite-horizon controller. Hence, we established the result that the infinite horizon optimal controller can be obtained as the limit of the finite-horizon one.

4   Derivation of Optimal Decentralized Flow Controllers for M Users

The results of the previous section can be extended to the most general case of M users to a certain extent. In derivations to follow, for ease of notation we will assume that one of the users has no delay in acquiring the queue length and effective service rate information from the switch. In other words, we let $D_1 = 0$ and delays of the other users can be taken to be ordered as in (2). As in the two user case, we use a dynamic programming approach to calculate the controls, but this time the expressions become too cumbersome to write down analytically. However, the main idea of our calculations remains essentially the same: completion of squares. Starting at stage N again, down to stage $N - D_2$, we minimize J_n's over the control of the user that has the least information delay, namely the user 1. The terms that remain after this minimization have the form

$$R_d = \sum_{n=0}^{N-d} \gamma_n \left[\sum_{l=2}^{M} \left(u_{l,N-n-D_2+1} - a_2 \xi_{N-n-D_2+1} \right) + \rho_{n,1} x_{N-D_2+1} \right.$$
$$\left. + \sum_{k=1, n \geq 1}^{n} \sum_{l=2}^{M} \rho_{n,k} \left(u_{l,N-D_2+k} - a_l \xi_{N-D_2+k} \right) \right]^2,$$
$$d = N, \ldots, N - D_2 + 1,$$

where γ_n's and $\rho_{n,k}$'s can be precomputed, and they depend on D_2 and the weights c_i's.

Between the stages $N - D_2$ and $N - D_3 + 1$, we minimize the objective functionals over two controls simultaneously. This brings in additional remaining terms on top of the slightly modified R_{N-D_2+1}. The new terms have the following structure:

$$A_d = \sum_{n=D_2}^{N-d} \gamma_n \left[\sum_{l=3}^{M} \left(u_{l,N-n-D_3+1} - a_2 \xi_{N-n-D_3+1} \right) + \rho_{n,1} x_{N-D_3+1} \right.$$
$$\left. + \sum_{k=1, n \geq 1}^{n} \sum_{l=3}^{M} \rho_{n,k} \left(u_{l,N-D_3+k} - a_l \xi_{N-D_3+k} \right) \right]^2,$$
$$d = N - D_2, \ldots, N - D_3 + 1, \tag{33}$$

and the modified R_{N-D_2+1} equals

$$\tilde{R}_{N-D_2+1} = \sum_{n=0}^{D_2-1} \gamma_n \left[\sum_{l=2}^{M} \left(u_{l,N-n-D_3+1} - a_2 \xi_{N-n-D_3+1} \right) + \rho_{n,1} x_{N-D_3+1} \right.$$
$$\left. + \sum_{k=1, n \geq 1}^{n} \sum_{l=2}^{M} \rho_{n,k} \left(u_{l,N-D_3+k} - a_l \xi_{N-D_3+k} \right) \right]^2. \tag{34}$$

Now, (33) and (34) can be combined to find all of the terms that are transferred to the next stage at time $N - D_3 + 1$. The result is

$$R_{N-D_3+1} = A_{N-D_3+1} + \tilde{R}_{N-D_2+1}.$$

From time $N - D_3$ until $N - D_4 + 1$, additional terms accumulate. These terms have the same structure as in (33) and are given by

$$
A_d = \sum_{n=D_3}^{N-d} \gamma_n \left[\sum_{l=4}^{M} \left(u_{l,N-n-D_4+1} - a_2 \xi_{N-n-D_4+1} \right) + \rho_{n,1} x_{N-D_4+1} \right.
$$

$$
\left. + \sum_{k=1, n \geq 1}^{n} \sum_{l=4}^{M} \rho_{n,k} \left(u_{l,N-D_4+k} - a_l \xi_{N-D_4+k} \right) \right]^2,
$$

$$
d = N - D_3, \ldots, N - D_4 + 1.
$$

Fortunately, the old terms that have already accumulated preserve their structure, but their time indices re-shift. Thus, $\hat{R}_{N-D_2+1}$ is re-modified to

$$
\hat{R}_{N-D_2+1} = \sum_{n=0}^{D_2-1} \gamma_n \left[\sum_{l=2}^{M} \left(u_{l,N-n-D_4+1} - a_2 \xi_{N-n-D_4+1} \right) + \rho_{n,1} x_{N-D_4+1} \right.
$$

$$
\left. + \sum_{k=1, n \geq 1}^{n} \sum_{l=2}^{M} \rho_{n,k} \left(u_{l,N-D_4+k} - a_l \xi_{N-D_4+k} \right) \right]^2
$$

and R_{N-D_3+1} becomes

$$
\tilde{R}_{N-D_3+1} = \sum_{n=D_2}^{D_3-1} \gamma_n \left[\sum_{l=3}^{M} \left(u_{l,N-n-D_4+1} - a_2 \xi_{N-n-D_4+1} \right) + \rho_{n,1} x_{N-D_4+1} \right.
$$

$$
\left. + \sum_{k=1, n \geq 1}^{n} \sum_{l=3}^{M} \rho_{n,k} \left(u_{l,N-D_4+k} - a_l \xi_{N-D_4+k} \right) \right]^2.
$$

We again combine all the terms to get

$$
R_{N-D_4+1} = A_{N-D_4+1} + \hat{R}_{N-D_2+1} + \tilde{R}_{N-D_3+1}.
$$

This process can be repeated several times till the stage $N - D_M$ is reached. Eventually, we find R_{N-D_M+1}. From that point on, this so-called transient minimization process ends, and we start minimizing J_n's over the controls of all of the M users. This task can be reduced to finding a recursive relation such as (13) in the two user case, but this time the algebra is too complicated to carry out the manipulations analytically by hand. However, a computer can do these calculations easily for us once we know that the structure of the remaining terms are as in the above expressions. This is actually what is done in the next section. Note that, the finite-horizon controller for the M user case is a certainty equivalent controller as in the two user case. This fact simply follows from the construction of the controller. Let us make a final remark on the infinite-horizon version of the M user problem. The only practical difficulty of the M user case is the complication of algebra, which prevents us from finding a recursive formula for the controller dynamics. We know that such a relation exists, but we do not know yet whether it is convergent or not. If we had to assume that it converges, the extension to the infinite-horizon case would be exactly the same as in the two user case. To put it another way, the infinite horizon optimal controller would be the limit of the finite horizon one.

5   Simulation Results

We performed an extensive simulation study to investigate the performance of the optimal controller. In all simulations, we considered a network of $M = 3$ users, with the following values of system parameters

$$a_1 = a_2 = a_3 = 1/3, \; c_1 = c_2 = c_3 = c, \; p = 2, \; \alpha_1 = \alpha_2 = 0.4, \; \sigma_\phi^2 = 1$$

where ϕ_n's are assumed to be $i.i.d$ Gaussian. And the run length of all simulations is $N = 1000$.

First, for comparison purposes, in Table 1, Table 2 and Table 3 we present the performance of the optimal controller along with two previously proposed certainty equivalent controllers referred to as Controller 1 and Controller 2 in [2]. As expected, optimal controller does a better job in regulating the queue length with comparable control effort and the resulting average cost is smaller in all cases. In [1] and [2], the simulation results indicated that the Controller 2 performs better than Controller 1, which is the case here, too. Next, we consider three scenarios to investigate the performance of the optimal controller only. In scenario 1, we decrease the delay of user 2 holding that of users 1 and 3 at their previous values. As depicted in Table 4, in this case the optimal overall cost decreases, because user 2 has access to more recent information about the queue length and available service rate. Next, we increase the delay of user 2 and study its effect on the optimal cost. Table 5 summarizes our simulation results in this case. Clearly, this time the optimal cost increases, because user 2 has worse knowledge on the system variables. Finally, we decrease the delay of user 3, while holding the delays of users 1 and 2 fixed at their base values. The performance of optimal controller in this case is depicted in Table 6. As the results indicate, the optimal average cost is smaller as compared to the case in which $D_3 = 10$. The decrease in the delay of user 3 results in a better performance of the system, because user 3 can make better estimates of the queue length and available service rate which are used in calculating his control actions and the overall cost.

Another issue we investigated in simulations is the rate of convergence of system dynamics. In all cases, the finite-horizon optimal control policies converged to stationary policies at very fast rates. This result suggests the existence of a stationary infinite horizon optimal controller for the general M user case, a result that we did not prove here.

Table 1. Optimal Controller with $D_1 = 0, D_2 = 5, D_3 = 10$.

c	$\sqrt{E(x^2)}$	$\sqrt{E(u_1^2)}$	$\sqrt{E(u_2^2)}$	$\sqrt{E(u_3^2)}$	J
0.1	0.0045	1.2830	0.2667	0.0898	0.0116
1	0.2178	1.3057	0.3271	0.1081	1.1545
10	3.1474	0.7896	0.3802	0.1773	66.7900

Table 2. Controller 1 with $D_1 = 0, D_2 = 5, D_3 = 10$.

c	$\sqrt{E(x^2)}$	$\sqrt{E(u_1^2)}$	$\sqrt{E(u_2^2)}$	$\sqrt{E(u_3^2)}$	J
0.1	2.69	0.89	0.63	0.49	7.68
1	3.18	0.84	0.60	0.49	11.44
10	7.19	0.38	0.48	0.50	114.66

Table 3. Controller 2 with $D_1 = 0, D_2 = 5, D_3 = 10$.

c	$\sqrt{E(x^2)}$	$\sqrt{E(u_1^2)}$	$\sqrt{E(u_2^2)}$	$\sqrt{E(u_3^2)}$	J
0.1	0.01	1.48	1.18	0.49	0.04
1	0.80	1.01	0.82	0.50	2.46
10	6.00	0.38	0.47	0.50	85.37

Table 4. Optimal Controller with $D_1 = 0, D_2 = 2, D_3 = 10$.

c	$\sqrt{E(x^2)}$	$\sqrt{E(u_1^2)}$	$\sqrt{E(u_2^2)}$	$\sqrt{E(u_3^2)}$	J
0.1	0.0042	1.1634	0.4776	0.0990	0.0085
1	0.2603	1.1012	0.5411	0.1049	0.8587
10	2.1791	0.6745	0.4920	0.1503	47.1239

Table 5. Optimal Controller with $D_1 = 0, D_2 = 8, D_3 = 10$.

c	$\sqrt{E(x^2)}$	$\sqrt{E(u_1^2)}$	$\sqrt{E(u_2^2)}$	$\sqrt{E(u_3^2)}$	J
0.1	0.0045	1.4276	0.1649	0.0978	0.0137
1	0.2013	1.4193	0.1919	0.1082	1.4320
10	3.9501	0.9362	0.3049	0.2215	90.6461

Table 6. Optimal Controller with $D_1 = 0, D_2 = 5, D_3 = 7$.

c	$\sqrt{E(x^2)}$	$\sqrt{E(u_1^2)}$	$\sqrt{E(u_2^2)}$	$\sqrt{E(u_3^2)}$	J
0.1	0.0048	1.4776	0.2902	0.1779	0.0140
1	0.2225	1.2763	0.2816	0.1571	1.1583
10	3.3677	0.8089	0.3747	0.2743	66.1541

6   Conclusions

In this paper, we have presented a method to calculate the optimal solution to a class of team problems with a partially nested information pattern, which particularly arises in the context of designing flow controllers in communication networks. As a basic approach towards obtaining the solution we adopted dynamic programming with which we first obtained the solution in a finite time horizon setting. Next, we exploited the special structure of the two user problem to extend this result to the infinite horizon case. We proved the existence of the infinite horizon solution and showed that it can be calculated as the limit of the finite horizon one. Further extension of this result to the M user case is still an open problem that we are currently working on. Yet another future direction of research would be to look into the continuous-time version of the same problem.

References

1. E. Altman, T. Başar, and R. Srikant, "A team-theoretic approach to congestion control," submitted to *Automatica*, June 1998.
2. E. Altman, T. Başar, and R. Srikant, "Robust rate control for ABR sources," In *Proceedings of IEEE INFOCOM*, San Francisco, CA, March 1998.
3. E. Altman, T. Başar, and R. Srikant, "Multi-user rate-based flow control with action delays: A team-theoretic approach.", In *Proceedings of the 36th IEEE Conference on Decision and Control*, pp. 2387–2392, San Diego, CA, December 1997.
4. E. Altman and T. Başar, "Optimal rate control for high speed telecommunication networks," In *Proceedings of the 34th IEEE Conference on Decision and Control*, pp. 1389–1394, New Orleans, LA, December 1995.
5. T. Başar and J. B. Cruz, Jr., "Concepts and methods in multi-person coordination and control," In S. G. Tzafestas, editor, *Optimization and Control of Dynamic Operational Research Models*, Chapter 11, pp. 351–387, North-Holland Publ. Co., 1982.
6. J-C. Bolot, "End-to-end packet delay and loss behavior in the Internet," in *Proceedings of ACM Sigcomm '93*, pp. 289–298, San Francisco, CA, September 1993.
7. Y. C. Ho and K. C. Chu, "Team decision theory and information structures in optimal control problems: Part I," *IEEE Transactions on Automatic Control*, vol. AC 17, pp. 15 21, 1972.
8. D. Kincaid and W. Cheney, *Numerical Analysis: Mathematics of Scientific Computing*, Brooks/Cole Publ. Co., 1991.
9. R. Radner, "Team decision problems," *Ann. Math. Statist.*, vol. 33, no. 3, pp. 857–881, 1962.
10. N. Sandell and M. Athans, "Solution of some non-classical LQG stochastic decision problems," *IEEE Transactions on Automatic Control*, vol. AC-19, pp. 108–116, 1974.

Solution of a Wavelet Crime *

Pu Qian and Bruce A. Francis

Department of Electrical and Computer Engineering, University of Toronto,
Toronto, Ontario, Canada M5S 1A4
Email: puqian, francis@control.utoronto.ca

Abstract. Wavelet theory is based on multiresolution. Applied to $\mathcal{L}^2$, this yields an infinite nest of approximation subspaces at decreasing scales of resolution. Wavelet subspaces are then formed from these approximation subspaces. The analysis of a signal $x(t)$ in $\mathcal{L}^2$ begins by projecting it onto one of the approximation subspaces. The pyramid algorithm then takes over for computing a lower scale approximation along with the detail signals. The input to the pyramid algorithm is the sequence of coefficients in the orthonormal basis of the approximation subspace. An interesting question is how to initialize the pyramid algorithm when only sampled values of $x(t)$ are available. Common practice is to use these sampled values directly in place of the correct coefficients. Strang and Nguyen charge this to be a "wavelet crime." This paper looks at the error incurred by this crime and proposes an optimal solution adapted from the theory of optimal sampled-data control.

1 Introduction

Wavelet theory is an important, current topic in signal processing and it has found many interesting applications, for example, to image compression. Following Daubechies' seminal paper [4], there are now several books on the subject, for example, [5], [7], [8], [9], as well as commercial software. Wavelet analysis is an alternative to Fourier analysis. Fourier analysis offers excellent, indeed optimal, resolution in the frequency domain, but poor resolution in the time domain. By this we mean that a short burst in the time domain has an infinite number of Fourier components. In wavelet theory frequency is replaced by the concept of *scale*, and wavelet analysis provides a compromise—good resolution with respect to both time and scale.

Our setting is the space $\mathcal{L}^2$ of continuous-time signals $x(t)$ defined for all t in $\mathbb{R}$. The most elegant way to develop wavelet theory is via *multiresolution analysis*. This yields a chain or nest of approximation subspaces at decreasing scales of resolution, each having an orthonormal basis. For each approximation subspace, its orthonormal basis is obtained by translating a kernel signal, called a *scaling function*; thus all basis signals at this scale have the same width of support in the time domain. This width is a measure of resolution—wider basis functions signify coarser resolution and narrower ones signify finer resolution. Wavelet subspaces are formed from these approximation subspaces by taking orthogonal complements.

* This work was supported by the Natural Sciences and Engineering Research Council (Canada).

The analysis of a signal begins by projecting it onto one of the approximation subspaces. The pyramid algorithm, implemented by a multirate filter bank, then takes over for computing a lower scale approximation along with the detail signals. The correct input to the pyramid algorithm is the sequence of coefficients in the orthonormal basis of the approximation subspace. An interesting question is how to initialize the pyramid algorithm when only sampled values of $x(t)$ are available. Common practice is to use these sampled values directly in place of the correct coefficients. Strang and Nguyen charge this to be a "wavelet crime." In this paper we look at the error incurred by this crime and propose an optimal solution, adapting techniques from the theory of optimal sampled-data control [6]. The importance of properly initializing the pyramid algorithm was also emphasized by Abry and Flandrin [1], but without an error analysis.

2   Scaling Function and Approximation Subspaces

We describe the rudiments of wavelet theory from scratch. The central concept is that of *multiresolution*, that is, representation at different scales of resolution. For the signal space $\mathcal{L}^2$, a *multiresolution structure* is a series $\{\mathcal{V}_m\}_{m\in\mathbb{Z}}$ of (closed) subspaces having the properties

1. They are nested:

$$\{0\} \subseteq \cdots \subseteq \mathcal{V}_1 \subseteq \mathcal{V}_0 \subseteq \mathcal{V}_{-1} \subseteq \cdots \subseteq \mathcal{L}^2.$$

2. The nest converges down to zero:

$$\lim_{m\to\infty} \mathcal{V}_m = \{0\}.$$

3. The nest converges up to $\mathcal{L}^2$:

$$\lim_{m\to-\infty} \mathcal{V}_m = \mathcal{L}^2.$$

We think of $\mathcal{V}_m$ as an *approximation subspace* of $\mathcal{L}^2$ at the scale of resolution indexed by m. Thus for $x(t)$ in $\mathcal{L}^2$, its orthogonal projection $v_m(t)$ onto $\mathcal{V}_m$ is viewed as the approximation of $x(t)$ at scale m. The scale of resolution increases as we move up the nest.

The simplest example to think about is the so-called *Haar multiresolution structure*. Here, $\mathcal{V}_0$ is the subspace of $\mathcal{L}^2$-signals that are constant on the intervals

$$\dots, [-2, -1), [-1, 0), [0, 1), [1, 2), \dots$$

of width 1. One step down the nest, $\mathcal{V}_1$ is the subspace of signals that are constant on the intervals

$$\dots, [-4, -2), [-2, 0), [0, 2), [2, 4), \dots$$

of width 2. In general, signals in $\mathcal{V}_m$ are constant on intervals of width 2^m. The nest structure is clear. The approximation $v_m(t)$ in $\mathcal{V}_m$ of an arbitrary $\mathcal{L}^2$-signal $x(t)$ is piecewise constant, the jumps spaced 2^m apart.

We pause at this point to bring in two operators: The *translation operator*

$$T : \mathcal{L}^2 \to \mathcal{L}^2, \quad (Tx)(t) = x(t-1)$$

and the *dilation operator*

$$D : \mathcal{L}^2 \to \mathcal{L}^2, \quad (Dx)(t) = \frac{1}{\sqrt{2}} \, x(t/2)$$

These operators have the following properties:

$$T \text{ and } D \text{ are unitary, } DT = T^2D, \; D^{-1}T = T^{1/2}D^{-1}.$$

In wavelet theory, a single function $\phi(t)$ in $\mathcal{L}^2$, known as the *scaling function*, is used to generate the entire multiresolution structure by the following construction: Introduce $\phi_{mn} := D^m T^n \phi$, or in the case $m = 0$ simply $\phi_n := T^n \phi$; then define

$$\mathcal{V}_m := \operatorname{span}_n \{\phi_{mn}\}.$$

The scaling function is designed to have two properties:

$$\{\phi_n\} \text{ are orthonormal;} \tag{1}$$

$$\phi \in \mathcal{V}_{-1}. \tag{2}$$

These two properties guarantee the nesting property of $\{\mathcal{V}_m\}$, and also that $\{\phi_{mn}\}_{n \in \mathbb{Z}}$ is an orthonormal basis for $\mathcal{V}_m$ for every m. In addition, it is customary to normalize ϕ so that

$$\int_{-\infty}^{\infty} \phi(t)dt = 1.$$

Property (2) is equivalent to the existence of coefficients h_n satisfying

$$\phi = D^{-1} \sum_n h_n \phi_n, \tag{3}$$

or more explicitly,

$$\phi(t) = \sqrt{2} \sum_n h_n \phi(2t - n).$$

This is a functional equation called the *dilation equation*. These coefficients form the impulse response of a discrete-time filter $H : \ell^2 \longrightarrow \ell^2$. It can be shown that there is a one-to-one correspondence between the scaling function

ϕ and the filter H. The filter H will later be very important in construction of the wavelet basis of $\mathcal{L}^2$, as well as in computing the wavelet transform in a technique known as the "pyramid algorithm."

For the Haar example,

$$\phi(t) = \begin{cases} 1, & 0 \le t < 1 \\ 0, & \text{otherwise} \end{cases}$$

and

$$h_n = \begin{cases} 1/\sqrt{2}, & n = 0, 1 \\ 0, & \text{otherwise.} \end{cases}$$

So H is the lowpass FIR filter with transfer function $(1 + z^{-1})/\sqrt{2}$.

Now we introduce half of the pyramid algorithm. Let $x(t)$ be an $\mathcal{L}^2$-signal and $v_m(t)$ its orthogonal projection in $\mathcal{V}_m$. Then v_m is a linear combination of the basis functions $\{\phi_{mn}\}_{n \in \mathbf{Z}}$:

$$v_m = \sum_n \tilde{v}_m(n)\phi_{mn}.$$

By orthonormality, the m-scale approximation coefficients are merely the inner products

$$\tilde{v}_m(n) = \langle \phi_{mn}, x \rangle.$$

For each m, $\tilde{v}_m \in \ell^2$. The pyramid algorithm is a procedure for recursively computing $\tilde{v}_m$. We have already defined the filter H, which mathematically is an LTI operator on ℓ^2. In addition, define the *downsampling operator*

$$S_\downarrow : \ell^2 \to \ell^2, \quad (S_\downarrow v)(n) = v(2n).$$

The pyramid algorithm states that

$$\tilde{v}_{m+1} = S_\downarrow H \tilde{v}_m.$$

The block diagram is the multirate DSP system in Fig. 1. Thus, by knowing

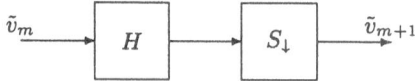

Fig. 1. Recursively computing the approximation coefficients.

the approximation coefficients, $\tilde{v}_m$, at scale m, one can compute them at the next scale down.

3   Wavelet Function and Wavelet Subspaces

Wavelet subspaces are defined in terms of a multiresolution structure as follows: $\mathcal{W}_m$ is the orthogonal complement of $\mathcal{V}_m$ in $\mathcal{V}_{m-1}$:

$$\mathcal{V}_{m-1} = \mathcal{V}_m \oplus \mathcal{W}_m.$$

Pictorially, the relationships among $\{\mathcal{V}_m\}$ and $\{\mathcal{W}_m\}$ can be represented by the lattice diagram in Fig. 2. As for approximation subspaces, we have $\mathcal{W}_m =$

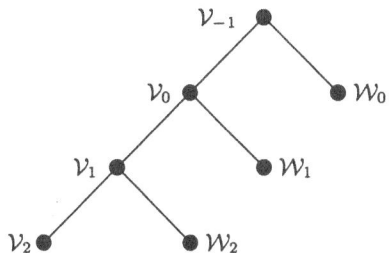

Fig. 2. Approximation subspaces and wavelet subspaces.

$D^m \mathcal{W}_0$. Also, the wavelet subspaces are pairwise orthogonal and

$$\mathcal{V}_M = \bigoplus_{m=M+1}^{\infty} \mathcal{W}_m.$$

Letting M tend to negative infinity, we get

$$\mathcal{L}^2 = \bigoplus_{m=-\infty}^{\infty} \mathcal{W}_m, \tag{4}$$

which states that $\{\mathcal{W}_m\}$ forms an orthogonal decomposition of $\mathcal{L}^2$.

For the Haar example, recall that $\mathcal{V}_0$ is the subspace of $\mathcal{L}^2$-signals that are constant on the intervals

$$\ldots, [-2, -1), [-1, 0), [0, 1), [1, 2), \ldots$$

of width 1 and $\mathcal{V}_1$ is the subspace of signals that are constant on the intervals

$$\ldots, [-4, -2), [-2, 0), [0, 2), [2, 4), \ldots$$

of width 2. So if $w(t)$ is in the orthogonal complement $\mathcal{W}_1$ of $\mathcal{V}_1$ in $\mathcal{V}_0$, then, for example, it equals some constant c on the first half of the interval $[0, 2)$ and $-c$ on the second half. Similarly on every other such interval of width 2. In this way a signal in $\mathcal{V}_0$ (i.e., constant on the intervals of width 1) is decomposed into its component in $\mathcal{V}_1$, which is its approximation at the lower scale, and its component in $\mathcal{W}_1$, which provides the missing detail.

Again, a single function, this time denoted $\psi(t)$, generates all the wavelet subspaces. It can be shown that there is an LTI operator G on ℓ^2 such that the operator

$$\begin{bmatrix} S_\downarrow H \\ S_\downarrow G \end{bmatrix} : \ell^2 \longrightarrow \ell^2 \oplus \ell^2$$

is unitary. Let g_n denote the impulse response of G. Then, analogous to (3), the *wavelet function* (or mother wavelet) is defined to be

$$\psi = D^{-1} \sum_n g_n \phi_n.$$

Moreover, denoting

$$\psi_n = T^n \psi, \quad \psi_{mn} = D^m T^n \psi,$$

we have that $\{\psi_{mn}\}_{n\in\mathbb{Z}}$ is an orthonormal basis for $\mathcal{W}_m$.

For the Haar example,

$$\psi(t) = \begin{cases} 1, & 0 \le t < 1/2 \\ -1, & 1/2 \le t < 1 \\ 0, & \text{otherwise} \end{cases}$$

and

$$g_n = \begin{cases} 1/\sqrt{2}, & n = 0 \\ -1/\sqrt{2}, & n = 1 \\ 0, & \text{otherwise.} \end{cases}$$

So G is the highpass FIR filter with transfer function $(1 - z^{-1})/\sqrt{2}$.

Now we can complete the pyramid algorithm. Again, let $x(t)$ be an $\mathcal{L}^2$-signal, $v_m(t)$ its orthogonal projection in $\mathcal{V}_m$, and $w_m(t)$ its orthogonal projection in $\mathcal{W}_m$. Then w_m is a linear combination of the basis functions $\{\psi_{mn}\}_{n\in\mathbb{Z}}$,

$$w_m = \sum_n \tilde{w}_m(n)\psi_{mn}, \tag{5}$$

where the wavelet coefficients, or detail coefficients, are

$$\tilde{w}_m(n) = \langle \psi_{mn}, x \rangle.$$

The pyramid algorithm states that

$$\tilde{w}_{m+1} = S_\downarrow G \tilde{v}_m$$

and the complete block diagram of the pyramid algorithm is in Fig. 3. This is iterated as many levels as desired.

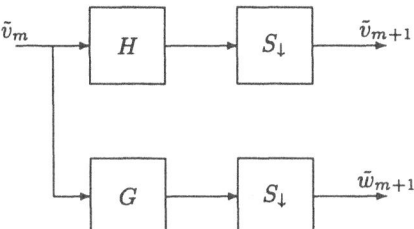

Fig. 3. The pyramid algorithm.

4   The Discrete Wavelet Transform

Equation (4) says that an arbitrary $x(t)$ can be decomposed into its wavelet components:

$$x = \sum_m w_m.$$

From (5), this expands to

$$x = \sum_m \sum_n \tilde{w}_m(n)\psi_{mn}.$$

Thus $\{\tilde{w}_m(n)\}$ are the coefficients in the expansion in the doubly-indexed wavelet basis $\{\psi_{mn}\}$. That is, $\tilde{w}_m(n)$ considered as a function of (m, n) is the transform of $x(t)$ with respect to this orthonormal basis. We therefore introduce a more suggestive notation $X(m, n) := \tilde{w}_m(n)$. The interpretation of the indices is

$$m = \text{scale}, \quad n = \text{time}.$$

This X is called the *discrete wavelet transform* (DWT) of x.

In summary, the *analysis equation* (going from x to X) is

$$X(m, n) = \langle \psi_{mn}, x \rangle$$

and the *synthesis equation* (going from X to x) is

$$x = \sum_m \sum_n X(m, n)\psi_{mn}.$$

The DWT has proven to be a wonderful way to represent an $\mathcal{L}^2$-signal, being able to provide good resolution in both the scale domain and the time domain.

5   The Initialization Problem

In practice, computation of the DWT for a signal $x(t)$ goes like this:

1. Select some level M where $v_M(t)$ in $\mathcal{V}_M$ represents $x(t)$ to a desired degree of resolution.
2. Get the approximation coefficients $\tilde{v}_M$ at this level.
3. Apply the pyramid algorithm to compute the approximation coefficients $\tilde{v}_{M+1}$ and the detail coefficients $\tilde{w}_{M+1}$ at the next scale down.
4. Repeat the pyramid algorithm a desired number of times, down to the coarsest desired scale.

So the pyramid algorithm is initialized by $\tilde{v}_M$. The correct initialization is therefore

$$\tilde{v}_M(n) = \langle \phi_{Mn}, x \rangle .$$

Let us study in detail how to compute this. Let F denote the continuous-time system with impulse response $2^{-M/2}\phi(-2^{-M}t)$. (So for $M = 0$, the impulse response of F is $\phi(-t)$.) We think of F as an LTI operator on $\mathcal{L}^2$. Also, let S_{2^M} denote the sampling operator of period 2^M, that is, S_{2^M} maps a continuous-time signal $f(t)$ to the discrete-time signal $f(2^M n)$. Then we have

$$
\begin{aligned}
\tilde{v}_M(n) &= \langle \phi_{Mn}, x \rangle \\
&= 2^{-M/2} \int_{-\infty}^{\infty} \phi(2^{-M}t - n)x(t)dt \\
&= 2^{-M/2} \int_{-\infty}^{\infty} \phi\left(\frac{t - 2^M n}{2^M}\right) x(t)dt \\
&= \text{convolution of } 2^{-M/2}\phi(-2^{-M}t) \text{ and } x(t) \text{ at time } 2^M n \\
&= (Fx)(2^M n) \\
&= (S_{2^M} Fx)(n)
\end{aligned}
$$

and hence $\tilde{v}_M = S_{2^M} Fx$. Thus $\tilde{v}_M(n)$ can be obtained from $x(t)$ by the signal processing in Fig. 4. Note that the entire continuous time function $x(t)$ is

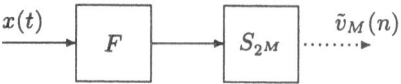

Fig. 4. Correct initialization of pyramid algorithm.

required in computing $\tilde{v}_M$.

We are interested in the case where one has only sampled measurements, so the available data are $\{x(\tau n)\}_{n\in\mathbb{Z}}$, where τ is the sampling period. Letting S_τ denote the sampling operator with period τ, the signal at hand is $S_\tau x$. Our problem is to compute $\tilde{v}_M$ from these samples. We propose a solution as illustrated in Fig. 5, where K_d is a suitably designed discrete-time linear

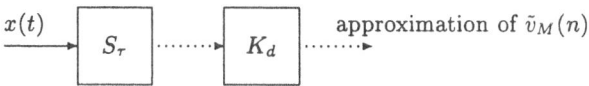

Fig. 5. Initialization from sampled data.

system, necessarily time varying in general if $2^M \neq \tau$. That is, we propose to initialize the pyramid algorithm by the sequence $K_d S_\tau x$.

It is not always possible to design K_d so that the pyramid algorithm is correctly initialized, that is, so that $K_d S_\tau x = S_{2^M} F x$. So it is natural to optimize K_d to minimize the ℓ^2-norm of the error

$$e = K_d S_\tau x - S_{2^M} F x = (K_d S_\tau - S_{2^M} F)x.$$

Moreover, it makes sense not to assume x is a fixed signal, but instead belongs to a class, and then $\|e\|_2$ is minimized for the worst x in its class. As in $\mathcal{H}^\infty$ control theory, we take the class to be the unit ball of $\mathcal{L}^2$ filtered by a lowpass continuous-time LTI weighting filter W on $\mathcal{L}^2$. Thus, the worst-case error norm is $\|(K_d S_\tau - S_{2^M} F)W\|$, the induced norm of the error operator

$$(K_d S_\tau - S_{2^M} F)W : \mathcal{L}^2 \longrightarrow \ell^2.$$

The task is to design K_d to minimize this induced error norm (IEN). The block diagram of the error system is shown in Fig. 6.

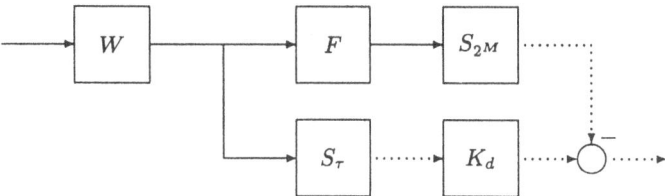

Fig. 6. The error system.

6 Optimal Filtering of the Sampled Data

In this section we study the optimization problem

$$\text{minimize}_{K_d}\|(K_dS_\tau - S_{2M}F)W\|. \tag{6}$$

For simplicity we shall do only the *special case* $M = 0$, $\tau = 1$. Then the two samplers are identical and we denote either simply by S. As mentioned before, in this special case F is the LTI operator with impulse response $\check{\phi}(t) := \phi(-t)$. Finally, since all the discrete-time signals in Fig. 6 are now at the same data rate, it is natural to restrict K_d to be LTI.

Problem (6) is a special case of a general sampled-data control problem [6], but here we give a more direct solution. Our solution is a formula for the optimal K_d in the frequency domain. We use ^ to denote discrete-time or continuous-time Fourier transform, so, for example, $\hat{k}_d(e^{j\omega})$ denotes the frequency response of K_d and $\hat{w}(j\omega)$ that of W.

Theorem 1. *Problem (6) has the (not necessarily unique) solution*

$$\hat{k}_d(e^{j\omega}) = \frac{\sum_k |\hat{w}(j\omega + j2\pi k)|^2 \hat{\phi}(-j\omega - j2\pi k)}{\sum_k |\hat{w}(j\omega + j2\pi k)|^2}.$$

The proof uses a preliminary result.

Lemma 1. *Let V be a continuous-time LTI system with frequency response $\hat{v}(j\omega)$ that rolls off fast enough that SV is bounded as an operator $\mathcal{L}^2 \longrightarrow \ell^2$. Then the induced norm of SV satisfies*

$$\|SV\|^2 = \sup_\omega \sum_k |\hat{v}(j\omega + j2\pi k)|^2.$$

Proof. Introduce the adjoint operator $(SV)^* : \ell^2 \longrightarrow \mathcal{L}^2$ of SV. Then

$$\|SV\|^2 = \|(SV)(SV)^*\|$$

and $(SV)(SV)^*$ is an LTI discrete-time operator on ℓ^2. Being LTI, its induced norm equals the supremal magnitude of its frequency response. So it remains to show that the frequency response of $(SV)(SV)^*$ equals

$$\sum_k |\hat{v}(j\omega + j2\pi k)|^2.$$

Let $\delta(n)$ denote the discrete-time unit impulse, let $v(t)$ denote the impulse response of V, and let $u(t)$ be an arbitrary signal in $\mathcal{L}^2$. Then

$$\langle (SV)u, \delta \rangle = \langle u, (SV)^*\delta \rangle.$$

The left-hand side equals

$$(v \star u)(0) = \int_{-\infty}^{\infty} v(-t)u(t)dt = \langle \check{v}, u \rangle,$$

where $\star$ denotes convolution and $\check{v}$ denotes the signal obtained from v by reversing time. Thus $(SV)^*\delta = \check{v}$. It follows that $V(SV)^*\delta$ equals $v \star \check{v}$, and then that the impulse response of $(SV)(SV)^*$ equals $S(v \star \check{v})$. From the action of the sampler S in the frequency domain, it is known that the discrete-time Fourier transform of $S(v \star \check{v})$ equals

$$\sum_k \widehat{(v \star \check{v})}(j\omega + j2\pi k).$$

Finally, $\widehat{(v \star \check{v})}(j\omega) = |\hat{v}(j\omega)|^2$. $\qquad\qquad\qquad\qquad\qquad\qquad\qquad$ $\Box$

Proof of Theorem 1 We first derive a frequency-domain formula for the square of the IEN in terms of the frequency responses of K_d, F, and W. Again, from the action of S in the frequency domain, the discrete-time Fourier transform of $K_d S x$ equals the discrete-time Fourier transform of SGx, where G is the continuous-time LTI system with frequency response $\hat{g}(j\omega) = \hat{k}_d(e^{j\omega})$. Thus, the initialization error becomes $e = S(G - F)Wx$, or $e = SVx$ where $V := (G - F)W$, and therefore the IEN is

$$\|(K_d S - SF)W\| = \|SV\|.$$

Now the frequency response of V is

$$\hat{v}(j\omega) = [\hat{g}(j\omega) - \hat{f}(j\omega)]\hat{w}(j\omega)$$
$$= [\hat{k}_d(e^{j\omega}) - \hat{\phi}(-j\omega)]\hat{w}(j\omega).$$

Combining these equations with the lemma gives the desired formula:

$$\|(K_d S - SF)W\|^2 - \sup_\omega \sum_k \left|[\hat{k}_d(e^{j\omega}) - \hat{\phi}(-j\omega - j2\pi k)]\hat{w}(j\omega + j2\pi k)\right|^2.$$
(7)

So problem (6) reduces to finding $\hat{k}_d(e^{j\omega})$ to minimize the right-hand side of (7).

Define

$$Q(\omega) = \sum_k \left|[\hat{k}_d(e^{j\omega}) - \hat{\phi}(-j\omega - j2\pi k)]\hat{w}(j\omega + j2\pi k)\right|^2$$

and

$$J = \sup_\omega Q(\omega).$$

We want to minimize J. Observe that $Q(\omega)$ is periodic of period 2π. Thus

$$J = \sup_{|\omega|<\pi} Q(\omega).$$

As will be seen, we can minimize $Q(\omega)$ at each value of ω; this is clearly optimal.

Write $\hat{k}_d$ and $\hat{\phi}$ in terms of their real and imaginary parts:

$$\hat{k}_d = \hat{k}_d^R + j\hat{k}_d^I, \quad \hat{\phi} = \hat{\phi}^R + j\hat{\phi}^I.$$

Then

$$Q(\omega) = \sum_k [\hat{k}_d^R(e^{j\omega}) - \hat{\phi}^R(-j\omega - j2\pi k)]^2 |\hat{w}(j\omega + j2\pi k)|^2$$
$$+ \sum_k [\hat{k}_d^I(e^{j\omega}) - \hat{\phi}^I(-j\omega - j2\pi k)]^2 |\hat{w}(j\omega + j2\pi k)|^2.$$

We can minimize the two right-hand terms independently. Let

$$Q_1(\omega) = \sum_k [\hat{k}_d^R(e^{j\omega}) - \hat{\phi}^R(-j\omega - j2\pi k)]^2 |\hat{w}(j\omega + j2\pi k)|^2.$$

Setting to 0 the first derivative of $Q_1(\omega)$ with respect to $\hat{k}_d^R(e^{j\omega})$ gives

$$\hat{k}_d^R(e^{j\omega}) = \frac{\sum_k |\hat{w}(j\omega + j2\pi k)|^2 \hat{\phi}^R(-j\omega - j2\pi k)}{\sum_k |\hat{w}(j\omega + j2\pi k)|^2}.$$

The second derivative is positive, so this $\hat{k}_d^R(e^{j\omega})$ is optimal. Similarly for $\hat{k}_d^I(e^{j\omega})$. Reassembling $\hat{k}_d(e^{j\omega})$ gives the desired result.   □

7   Example

This example illustrates how injurious the wavelet crime can be. The underlying wavelet is taken to be the Haar wavelet for simplicity. Since the sampling frequency is 1 Hz, there will be no aliasing provided the bandwidth of $x(t)$ is less than 0.5 Hz. To be more realistic, we relax this and allow some small energy at higher frequencies. We do this by choosing a non-ideal lowpass weighting filter W with cutoff 0.5 Hz, say the one in Fig. 7.

We consider two choices for the initialization filter K_d: the wavelet crime filter, $\hat{k}_d(e^{j\omega}) = 1$; the optimal filter given by Theorem 1. Unfortunately, the optimal filter has an irrational transfer function, so an FIR approximation would be needed in practice. The IEN $\|(K_dS - SF)W\|$ for the two cases is computed using the formula in Lemma 1. Since absolute numbers are not so meaningful, we normalize by $\|SFW\|$, which is the IEN for $K_d = 0$, the filter that sets to zero all the sampled data. The results are as follows:

	$\|(K_dS - SF)W\|/\|SFW\|$
wavelet crime filter	1.4399
optimal filter	0.0340

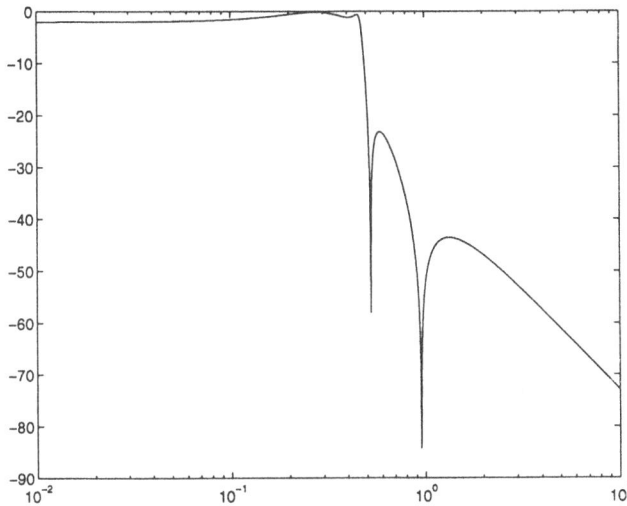

Fig. 7. Weighting filter frequency response: gain in dB versus frequency in Hz.

So the wavelet crime filter $(K_d = I)$ is actually worse than setting all the sampled data to zero $(K_d = 0)$. The optimal filter is an improvement by a factor of 42.

Let us emphasize that the problem so far is for the special case where the initial approximation space is $\mathcal{V}_0$ and the sampling frequency is 1 Hz. More generally, the problem has three parameters: the scale M of the initial approximation space $\mathcal{V}_M$, the frequency at which the signal $x(t)$ was sampled, and the bandwidth of W, which in turn reflects the bandwidth of $x(t)$. A more complete study would show the effects of these three parameters on the IEN.

We will obtain some preliminary results by allowing the bandwidth of W to be a variable. We do this by defining $W_N(j\omega) = W(j2^N\omega), N \geq 1$. Thus as N increases, the bandwidth of W_N decreases. The problem again is to minimize $\|(K_dS - SF)W_N\|$. Fig. 8 shows plots of normalized IEN versus N for the two cases. As N increases, both errors decrease, as they must: The wavelet crime filter curve decreases relatively slowly; the optimal filter curve shows a more rapid improvement. The left-hand limits $(N = 0)$ are the values in the previous table.

8   Conclusion

Is there a link between this paper and the work of Davison, in whose honor we write this? No. However, inspired by Davison's work on large-scale systems, it would be interesting to attempt to develop a theory of multiresolution of *systems* just as wavelet theory is based on multiresolution of *signals*. Along this line is the work of Willsky and co-workers [2], [3].

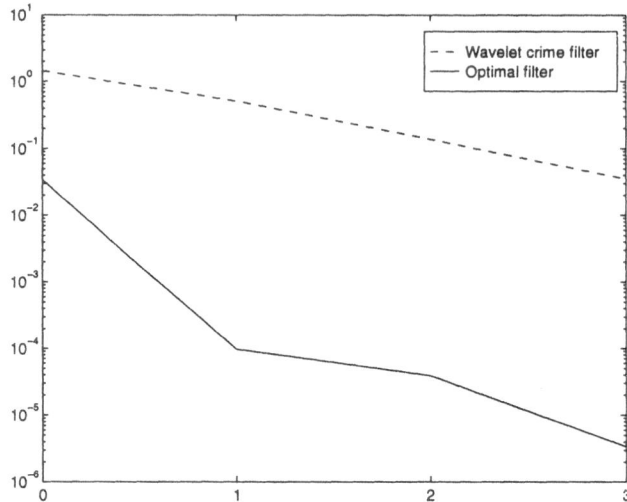

Fig. 8. Normalized induced error norm versus bandwidth parameter N.

References

1. P. Abry and P. Flandrin, "On the initialization of the discrete wavelet transform algorithm," *IEEE Signal Processing Letters*, vol. 1, no. 2, pp. 32–34, 1994.
2. A. Benveniste, R. Nikoukhah, and A. Willsky. "Multiscale system theory," *IEEE Trans. Circuits and Systems, Part I*, vol 41, pp. 2–15, 1994.
3. K. Chou, A. Willsky, and R. Nikoukhah, "Multiscale systems, Kalman filters, and Riccati equations," *IEEE Trans. Auto. Control*, vol. 39, pp. 479–492, 1994.
4. I. Daubechies, "Orthonormal bases of compactly supported wavelets," *Comm. Pure Appl. Math.*, vol. 41, pp. 909–996, 1988.
5. I. Daubechies, *Ten Lectures on Wavelets*, SIAM, Philadelphia, 1992.
6. T. Chen and B. Francis, *Optimal Sampled-Data Control Systems*, Springer-Verlag, Berlin, 1995.
7. Y. Meyer, *Wavelets: Algorithms and Applications*, SIAM, Philadelphia, 1993.
8. G. Strang and T. Nguyen, *Wavelets and Filter Banks*, Wellesley-Cambridge Press, Wellesley, MA, 1996.
9. M. Vetterli and J. Kovacevic, *Wavelets and Subband Coding*, Prentice-Hall, Englewood Cliffs, 1995.

Interior and Non-Interior Point Algorithms for Nonlinear Programming

Victor H. Quintana[1] and Geraldo L. Torres[2]

[1] University of Waterloo, Department of Electrical and Computer Engineering, 200 University Ave West, N2L 3G1, Waterloo, Ontario, Canada
Email: quintana@kingcong.uwaterloo.ca
[2] Universidade Federal de Pernambuco, Departamento de Engenharia Elétrica e Sistemas de Potência, Recife, Pernambuco, Brazil
Email: gltorres@kingcong.uwaterloo.ca

Abstract. The solution of an *optimal power flow* (OPF) problem by *interior-point* (IP) and *non-interior-point* (NIP) methods for *nonlinear programming* (NLP) is described. The IP and NIP algorithms are derived in details from a general NLP problem model that is suitable to express most OPF problems. The NIP algorithm, in contrast to IP algorithms, can start from arbitrary points. Numerical results illustrate the viability of the proposed algorithms as applied to several power networks that range in size from 30 to 2098 nodes.

1 Introduction

To correct undesirable operation conditions, power system operators are required to constantly set the control reference signal of each generator. This is done by controlling the production, absorption, and flow of power at all levels in the power system. The control signals are the generator outputs, transformer tap settings, shunt var sources, and so forth. Deciding on an optimal control signal, aiming at the secure and economic operation of a power system, is an extremely difficult task, which is best performed by the *optimal power flow* (OPF) tool at power system control centers. The OPF tool is a sophisticated computational procedure that uses optimization techniques to find an optimal setting of the system control variables, subject to a reasonably large set of specified physical and operational constraints.

Optimization of power system operations is one of the areas where *interior-point* (IP) methods are being applied extensively [1]. The power-engineering problems already solved by primal-dual IP methods include state estimation [2] and a variety of OPF problems [3–5], including minimum load shedding [6] and maximum loadability [7] of power systems, which are highly nonlinear problems of the OPF class. In the works reported in [2–7], very large *nonlinear programming* (NLP) problems have efficiently been solved using IP methods for NLP, which are derived from the logarithmic barrier function approach [8].

A typical NLP problem that frequently arises in power-engineering has the following general mathematical formulation:

$$\begin{aligned}
\text{minimize} \quad & f(x) \\
\text{subject to} \quad & g(x) = 0, \\
& \underline{h} \le h(x) \le \overline{h},
\end{aligned} \tag{1}$$

where, in a typical *optimal power flow* (OPF) problem,

- $x \in \mathbb{R}^n$ is a vector of decision variables, including the control and the nonfunctional dependent variables.
- $f : \mathbb{R}^n \to \mathbb{R}$ is a scalar function that represents the power system's operation optimization goal.
- $g : \mathbb{R}^n \to \mathbb{R}^m$ is a vector function with conventional power flow equations and other equality constraints.
- $h : \mathbb{R}^n \to \mathbb{R}^p$ is a vector of nonlinear functional bound and simple bound variables, with lower bound $\underline{h}$ and upper bound $\overline{h}$, corresponding to physical and operating limits on the system.

We assume that $f(x)$, $g_i(x)$ and $h_i(x)$ are twice continuously differentiable functions in $\mathbb{R}^n$.

The nonlinear problem (1) can be solved by primal-dual IP methods in two ways: (i) by applying these methods directly to the nonlinear problem, or (ii) by applying them to a sequence of (local) approximations of the problem, as in the *sequential linear programming* and *sequential quadratic programming* approaches. In the next section, we describe the mathematical development of a primal-dual IP method for solving (1) in a nonlinear manner.

In this paper, a *non-interior-point* (NIP) method for solving (1) is also proposed. It handles the complementarity conditions by means of a function $\psi : \mathbb{R}^2 \to \mathbb{R}$ that satisfies the property

$$\psi(a, b) = 0 \quad \Longleftrightarrow \quad a \ge 0, \quad b \ge 0, \quad ab = 0. \tag{2}$$

Since the non-negativity of any limit point is automatically assured by $\psi(\cdot, \cdot)$, the iterates do not necessarily have to stay in the positive orthant, in contrast to IP methods, and the algorithm can start from arbitrary points.

The paper is organized as follows. In Sect. 2 we present a detailed derivation of the IP method from the NLP problem (1). A *predictor-corrector* variant of this IP method is described in Sect. 3. In Sect. 4 we introduce the NIP algorithm. Numerical results are reported in Sect. 5. Some conclusions in Sect. 6 close the paper.

2   Interior-Point Method for Nonlinear Programming

The IP algorithm to solve (1) operates on a modified problem that emerges as we transform all inequalities in (1) into equalities by adding non-negative slack vectors, $(s, z) \in \mathbb{R}_+^p \times \mathbb{R}_+^p$, which are in turn incorporated into logarithmic barrier terms appended to the objective function, as follows:

$$
\begin{aligned}
\text{minimize} \quad & f(x) - \mu^k \sum_{i=1}^{p} (\ln s_i + \ln z_i) \\
\text{subject to} \quad & g(x) = 0, \\
& -s - z + \overline{h} - \underline{h} = 0, \\
& -h(x) - z + \overline{h} = 0, \quad s > 0, \ z > 0,
\end{aligned}
\tag{3}
$$

where $\mu^k > 0$ is a *barrier parameter* that is forced to decrease to zero as iterations progress. The strict positivity conditions on the slacks, imposed by the logarithm functions, are handled in an implicit manner. The sequence $\{\mu^k\}$ generates a sequence of sub-problems that are defined by (3); under *regularity assumptions* (see [9]), as μ^k approaches zero the sequence $\{x(\mu^k)\}$ of solutions of (3) approaches x^*, a local minimizer of (1).

To solve the equality-constrained problem (3), we use a Lagrange-Newton method. Towards this purpose, we define the following *Lagrangian function*:

$$
\begin{aligned}
\mathcal{L}(y; \mu^k) := & f(x) - \mu^k \sum_{i=1}^{p} \left(\ln s_i + \ln z_i \right) - \lambda^T g(x) - \pi^T \left(- s - z + \overline{h} - \underline{h} \right) \\
& - v^T \left(- h(x) - z + \overline{h} \right)
\end{aligned}
\tag{4}
$$

where $(\lambda, \pi, v) \in \mathbb{R}^m \times \mathbb{R}^p \times \mathbb{R}^p$ are vectors of *Lagrange multipliers*, called *dual variables*, and $y \in \mathbb{R}^q$ is the vector $y := (s, z, \pi, v, x, \lambda)^T$. A local minimizer of (3) is expressed in terms of a stationary point of $\mathcal{L}(y; \mu^k)$, which in turn must satisfy the first-order *Karush-Kuhn-Tucker* (KKT) conditions

$$
\nabla_y \mathcal{L}(y; \mu^k) =
\begin{pmatrix}
S\pi - \mu^k e \\
Z\widehat{v} - \mu^k e \\
s + z - \overline{h} + \underline{h} \\
h(x) + z - \overline{h} \\
\nabla_x f(x) - G(x)\lambda + H(x)v \\
-g(x)
\end{pmatrix}
= 0
\tag{5}
$$

where $\nabla_x f : \mathbb{R}^n \rightarrow \mathbb{R}^n$ is the gradient of $f(x)$, $G : \mathbb{R}^n \rightarrow \mathbb{R}^{n \times m}$ is the transposed Jacobian of $g(x)$, $H : \mathbb{R}^n \rightarrow \mathbb{R}^{n \times p}$ is the transposed Jacobian of $h(x)$, $S := \text{diag}(s_1, \dots, s_p)$, $Z := \text{diag}(z_1, \dots, z_p)$, $e := (1, 1, \dots, 1)^T$, and, for the sake of presentation, we define the vector $\widehat{v} := v + \pi$.

The KKT system (5) may be interpreted as follows. The third, the fourth, and the sixth terms in (5), along with $(s, z) \geq 0$, ensure *primal feasibility*; the fifth term, together with $(\pi, \widehat{v}) \geq 0$, ensure *dual feasibility*; while the first and the second terms are called the μ-*complementarity conditions*, perturbations $(\mu^k \neq 0)$ of the standard complementarity conditions.

Primal-dual IP iterates invariably apply *one step* of Newton's method for nonlinear equations to the KKT system (5), compute a step size in the Newton direction, update the variables, and reduce μ^k. The algorithm terminates when primal and dual infeasibility, and the complementarity gap fall below predetermined tolerances. The major steps of the algorithm are detailed below.

2.1 Solving for the Newton Directions

Although the KKT system (5) is nonlinear, its solution is usually approximated by a single iteration of Newton's method (the Newton direction is only a means to follow the path of minimizers parameterized by μ^k). As we apply Newton's method to solve (5), we obtain the following indefinite system:

$$
\begin{bmatrix}
\Pi & 0 & S & 0 & 0 & 0 \\
0 & \widehat{\Upsilon} & Z & Z & 0 & 0 \\
I & I & 0 & 0 & 0 & 0 \\
0 & I & 0 & 0 & H^T & 0 \\
0 & 0 & 0 & H & \nabla_x^2\mathcal{L} & -G \\
0 & 0 & 0 & 0 & -G^T & 0
\end{bmatrix}
\begin{pmatrix}
\Delta s \\
\Delta z \\
\Delta \pi \\
\Delta v \\
\Delta x \\
\Delta \lambda
\end{pmatrix}
= -
\begin{pmatrix}
S\pi - \mu^k e \\
Z\widehat{v} - \mu^k e \\
s + z - \overline{h} + \underline{h} \\
h(x) + z - \overline{h} \\
\nabla_x f(x) - G(x)\lambda + H(x)v \\
-g(x)
\end{pmatrix}
\tag{6}
$$

where $\Pi := \mathrm{diag}(\pi_1,\dots,\pi_p)$ and $\widehat{\Upsilon} := \mathrm{diag}(\widehat{v}_1,\dots,\widehat{v}_p)$; we have dropped most superscripts k to simplify the notation. The computation of the Hessian $\nabla_x^2\mathcal{L}$ involves a combination of the objective function Hessian $\nabla_x^2 f(x)$ and the constraint Hessians $\nabla_x^2 g_j(x)$ and $\nabla_x^2 h_j(x)$, as follows:

$$
\nabla_x^2\mathcal{L}(y) = \nabla_x^2 f(x) - \sum_{j=1}^{m} \lambda_j \nabla_x^2 g_j(x) + \sum_{j=1}^{p} v_j \nabla_x^2 h_j(x).
\tag{7}
$$

2.2 Updating the Variables

New values of the primal and dual variables are computed from

$$
x^{k+1} := x^k + \alpha_P^k \Delta x, \tag{8a}
$$
$$
s^{k+1} := s^k + \alpha_P^k \Delta s, \tag{8b}
$$
$$
z^{k+1} := z^k + \alpha_P^k \Delta z, \tag{8c}
$$
$$
\lambda^{k+1} := \lambda^k + \alpha_D^k \Delta \lambda, \tag{8d}
$$
$$
\pi^{k+1} := \pi^k + \alpha_D^k \Delta \pi, \tag{8e}
$$
$$
v^{k+1} := v^k + \alpha_D^k \Delta v, \tag{8f}
$$

where the scalars $\alpha_P^k \in (0,1]$ and $\alpha_D^k \in (0,1]$ are the *step length* parameters in primal and dual spaces, respectively. The maximum step lengths that can be taken along Δy are determined by

$$
\alpha_P^k := \min\left\{ 1,\ \gamma \min_i \left\{ \frac{-s_i^k}{\Delta s_i}\Big| \Delta s_i < 0,\ \frac{-z_i^k}{\Delta z_i}\Big| \Delta z_i < 0 \right\} \right\},
\tag{9a}
$$
$$
\alpha_D^k := \min\left\{ 1,\ \gamma \min_i \left\{ \frac{-\pi_i^k}{\Delta \pi_i}\Big| \Delta \pi_i < 0,\ \frac{-\widehat{v}_i^k}{\Delta \widehat{v}_i}\Big| \Delta \widehat{v}_i < 0 \right\} \right\}.
\tag{9b}
$$

The scalar $\gamma \in (0,1)$ is a *safety factor* to ensure that the next point will satisfy the strict positivity conditions; a typical value is $\gamma = 0.99995$.

Separate step lengths in the primal and dual spaces, as presented above, has proven highly efficient in LP. For general NLP, however, the interdependence of primal and dual variables in the dual feasibility equations does not rigorously allows for separate step lengths be taken in the primal and dual spaces. In such a case, a single common step length might be computed from

$$\alpha_P^k = \alpha_D^k \leftarrow \min\{\alpha_P^k, \alpha_D^k\}. \tag{10}$$

2.3   Reducing the Barrier Parameter

The residual of the complementarity conditions, called the *complementarity gap* ρ^k, is computed at the point y^k from

$$\rho^k = (s^k)^T \pi^k + (z^k)^T \widehat{v}^k. \tag{11}$$

The sequence $\{\rho^k\} \to 0$ provided that $\{x(\mu^k)\} \to x^*$. Thus, the relationship between ρ^k and μ^k that is implicit in the KKT equations (5) suggests that μ^k could be reduced based on a predicted decrease of ρ^k. For instance, as

$$\mu^{k+1} := \sigma \frac{\rho^k}{2p} \tag{12}$$

where $\sigma \in (0,1]$ is the expected, but not necessarily realized, decrease in the complementarity gap; we take $\sigma = 0.2$.

2.4   Testing Convergence

The IP iterations are considered terminated when the criteria

$$\nu_1^k \leq \epsilon_1, \tag{13a}$$
$$\nu_2^k \leq \epsilon_1, \tag{13b}$$
$$\nu_3^k \leq \epsilon_2, \tag{13c}$$
$$\nu_4^k \leq \epsilon_2, \tag{13d}$$

are satisfied, where

$$\nu_1^k := \max\left\{\max\{\underline{h} - h(x^k)\}, \max\{h(x^k) - \overline{h}\}, \|g(x^k)\|_\infty\right\}, \tag{14a}$$

$$\nu_2^k := \frac{\|\nabla_x f(x^k) - G(x^k)\lambda^k + H(x^k)v^k\|_\infty}{1 + \|x^k\| + \|\lambda^k\| + \|v^k\|}, \tag{14b}$$

$$\nu_3^k := \frac{\rho^k}{1 + \|x^k\|}, \tag{14c}$$

$$\nu_4^k := \frac{|f(x^k) - f(x^{k-1})|}{1 + |f(x^k)|}. \tag{14d}$$

The max-operator, as applied to vectors, is to be interpreted componentwise. If the criteria $\nu_1^k \leq \epsilon_1$, $\nu_2^k \leq \epsilon_1$ and $\nu_3^k \leq \epsilon_2$ are satisfied, then the point y^k is a KKT point of accuracy ϵ_1. When numerical problems prevent verifying this condition, the algorithm stops as soon as feasibility of the equality constraints is (hopefully) achieved along with very small fractional change in the objective value and negligible changes in the variables. Typical convergence tolerances are $\epsilon_1 = 10^{-4}$, $\epsilon_2 = 10^{-2}\epsilon_1$, and $\epsilon_\mu = 10^{-12}$.

An outline of the primal-dual IP algorithm is as follows:

Step 0: *(Initialization)*
Set $k = 0$, define μ^0 and choose a y^0 that satisfy the strict positivity conditions.

Step 1: *(Compute the Newton direction)*
Form the Newton system at the current point and solve for the direction Δy.

Step 2: *(Update Variables)*
Compute the step length α^k in the direction Δy and update primal and dual variables.

Step 3: *(Test Convergence and Update μ)*
If the new point satisfies convergence criteria, stop. Otherwise, set $k = k + 1$, compute μ^k and go back to **Step 1**.

3   Predictor-Corrector Interior-Point Algorithm

The predictor-corrector method developed by Mehrotra [10] has shown, in computational practice, to considerably improve the performance of primal-dual IP methods for LP. Rather than applying Newton's method to (5) to obtain correction terms to the estimate y^k, the new point $y^{k+1} = y^k + \Delta y$ is directly substituted into (5), to obtain

$$
\nabla_y^2 \mathcal{L}(y)
\begin{pmatrix}
\Delta s \\
\Delta z \\
\Delta \pi \\
\Delta v \\
\Delta x \\
\Delta \lambda
\end{pmatrix}
= -
\begin{pmatrix}
S\pi - \mu^k e + \Delta S \Delta \pi \\
Z\hat{v} - \mu^k e + \Delta Z \Delta \hat{v} \\
s + z - \overline{h} + \underline{h} \\
h(x) + z - \overline{h} \\
\nabla_x f(x) - G(x)\lambda + H(x)v \\
-g(x)
\end{pmatrix}
\tag{15}
$$

where $\nabla_y^2 \mathcal{L}(y)$ is the same coefficient matrix in the Newton system (6), $\Delta S := \mathrm{diag}(\Delta s_1, \dots, \Delta s_p)$, and $\Delta Z := \mathrm{diag}(\Delta z_1, \dots, \Delta z_p)$.

The major difference between the Newton systems (15) and (6) is that the right-hand side of (15) cannot be determined beforehand because of the non-linear *delta terms*. To solve (15), Mehrotra suggests first dropping the μ *terms* and the *delta terms* in the right-hand side of (15) and, then, solving for the (pure Newton) *affine-scaling direction*

$$
\nabla_y^2 \mathcal{L}(y)
\begin{pmatrix}
\Delta s_{\mathrm{af}} \\
\Delta z_{\mathrm{af}} \\
\Delta \pi_{\mathrm{af}} \\
\Delta v_{\mathrm{af}} \\
\Delta x_{\mathrm{af}} \\
\Delta \lambda_{\mathrm{af}}
\end{pmatrix}
= -
\begin{pmatrix}
S\pi \\
Z\hat{v} \\
s + z - \overline{h} + \underline{h} \\
h(x) + z - \overline{h} \\
\nabla_x f(x) - G(x)\lambda + H(x)v \\
-g(x)
\end{pmatrix} .
\tag{16}
$$

The direction Δy_{af} is then used in two distinct ways: (i) to approximate the *delta terms* in the right-hand side of (15), and (ii) to dynamically estimate the barrier parameter μ.

To estimate μ, first we consider the ratio test (10) to determine the step that would actually be taken if the affine direction Δy_{af} were used. Second, an estimate of the complementarity gap is computed from

$$
\rho_{\mathrm{af}}^k = (s^k + \alpha_P^{\mathrm{af}} \Delta s_{\mathrm{af}})^T (\pi^k + \alpha_D^{\mathrm{af}} \Delta \pi_{\mathrm{af}}) + (z^k + \alpha_P^{\mathrm{af}} \Delta z_{\mathrm{af}})^T (\hat{v}^k + \alpha_D^{\mathrm{af}} \Delta \hat{v}_{\mathrm{af}}).
\tag{17}
$$

Finally, an estimate μ_{af}^k for μ^{k+1} may be obtained from

$$
\mu_{\mathrm{af}}^k := \min \left\{ \left(\frac{\rho_{\mathrm{af}}^k{}^2}{\rho^k} \right),\ 0.2 \right\} \frac{\rho_{\mathrm{af}}^k}{2p} .
\tag{18}
$$

This chooses μ_{af}^k to be small when the affine direction produces a large decrease in complementarity, and chooses μ_{af}^k to be large otherwise. The actual new steps are computed from

$$
\nabla_y^2 \mathcal{L}(y)
\begin{pmatrix}
\Delta s \\
\Delta z \\
\Delta \pi \\
\Delta v \\
\Delta x \\
\Delta \lambda
\end{pmatrix}
= -
\begin{pmatrix}
S\pi - \mu^{\mathrm{af}} e + \Delta S^{\mathrm{af}} \Delta \pi^{\mathrm{af}} \\
Z\hat{v} - \mu^{\mathrm{af}} e + \Delta Z^{\mathrm{af}} \Delta \hat{v}^{\mathrm{af}} \\
s + z - \overline{h} + \underline{h} \\
h(x) + z - \overline{h} \\
\nabla_x f(x) - G(x)\lambda + H(x)v \\
-g(x)
\end{pmatrix} .
\tag{19}
$$

The additional effort in the predictor-corrector method is in the extra linear system solution to compute the affine direction, and the extra ratio test used to compute μ_{af}^k, since the predictor and corrector steps are based on the same matrix factorization. What is usually gained is reduction in iterations and solution time.

4   Non-Interior Point Method

In this section, we describe a non-interior-point approach for solving the NLP problem (1). This approach handles the complementarity conditions—$s_i \geq 0$,

$\pi_i \geq 0$, and $s_i \pi_i = 0$; $z_i \geq 0$, $\hat{v}_i \geq 0$ and $z_i \hat{v}_i = 0$—by means of a function $\psi_\mu : \mathbb{R}^2 \to \mathbb{R}$ which is defined by the following characterization of its zeros:

$$\psi_\mu(a, b) = 0 \quad \Longleftrightarrow \quad a > 0, \quad b > 0, \quad ab = \mu, \tag{20}$$

for any $\mu > 0$. Any function $\psi_\mu : \mathbb{R}^2 \to \mathbb{R}$ that have the property (20) is called an *NCP-function*. Several NCP-functions have been identified in the past few years [11,12]. Deserving special attention is the NCP-function

$$\psi_\mu(a, b) := a + b - \sqrt{a^2 + b^2 + 2\mu}, \tag{21}$$

that is employed by Chen and Harker [11] and Kanzow [12] to solve the *linear complementarity problem* LCP(q, M): given $M \in \mathbb{R}^{n \times n}$ and $q \in \mathbb{R}^n$, find $x \in \mathbb{R}^n$ and $y \in \mathbb{R}^n$ so that

$$Mx + q = y, \quad x \geq 0, \quad y \geq 0, \quad x^T y = 0. \tag{22}$$

By using NCP-functions, the Chen–Harker [11] and Kanzow [12] algorithms solve LCP(q, M) by *approximately* solving a sequence of nonlinear systems of equations of the form

$$\Psi_\mu(x) := \begin{pmatrix} Mx + q - y \\ \psi_\mu(x_1, y_1) \\ \psi_\mu(x_2, y_2) \\ \vdots \\ \psi_\mu(x_n, y_n) \end{pmatrix} = 0, \tag{23}$$

where $\mu > 0$ is a *continuation parameter* that is forced to decrease to zero, as does the barrier parameter in an IP setting.

Since the non-negativity of any limit point is automatically assured by an NCP-function, the attraction domain of the feasible path in the non-interior continuation method is all of $\mathbb{R}^n$ instead of the positive orthant $\mathbb{R}^n_+$, in contrast to IP methods. Then, the iterates do not necessarily have to stay in the positive orthant, and the algorithm can start with an arbitrary vector (x^0, y^0) which satisfies the condition $Mx^0 + q = y^0$.

A similar approach has been employed in the solution of the *nonlinear complementarity problem* (NCP) (see [13]), which is to find an $x \in \mathbb{R}^n$ so that

$$x \geq 0, \quad F(x) \geq 0, \quad x^T F(x) = 0, \tag{24}$$

where $F : \mathbb{R}^n \to \mathbb{R}^n$ is a vector of continuously differentiable functions. The reformulation of NCP(F), as a parametrized sequence of nonlinear systems of equations, has the form

$$\Psi_\mu(x) := \begin{pmatrix} \psi_\mu(x_1, F_1(x)) \\ \psi_\mu(x_2, F_2(x)) \\ \vdots \\ \psi_\mu(x_n, F_n(x)) \end{pmatrix} = 0. \tag{25}$$

The systems of equations (23) and (25) are usually solved by Newton-type methods, which, however, are in general only locally convergent. To globalize the local method, an Armijo-type line search is performed to minimize a *merit function*, usually $\Phi_\mu(x) := \frac{1}{2}\Psi_\mu(x)^T\Psi_\mu(x)$. The global minimizer of $\Phi_\mu(x)$ is a solution to $\Psi_\mu(x) = 0$.

Besides having the property (20), the NCP-function (21) is continuously differentiable for all $(a, b) \in \mathbb{R}^2$, and its partial derivatives have the property

$$\left.\frac{\partial\psi_\mu}{\partial a}\right|_{(a,b)} \in (0,2) \quad \text{and} \quad \left.\frac{\partial\psi_\mu}{\partial b}\right|_{(a,b)} \in (0,2) \tag{26}$$

for all $(a, b) \in \mathbb{R}^2$. Furthermore, Burke and Xu [14, Lemma 4.5] show that for any $\epsilon > 0$, if $\|\psi_\mu(x, y)\|_\infty \le \epsilon$, then

$$-\epsilon \le x_i, \quad -\epsilon \le y_i, \quad \text{and} \quad \frac{|x_iy_i - \mu|}{|x_i| + |y_i| + \sqrt{\mu}} \le \epsilon. \tag{27}$$

Yet, if $\mu \le \epsilon$ and $\|\psi_\mu(x, y)\|_2 \le \beta\mu$, then

$$-\sqrt{\beta\epsilon} \le x_i, \quad -\sqrt{\beta\epsilon} \le y_i, \quad \text{and} \quad \frac{|x_iy_i - \mu|}{|x_i| + |y_i| + \sqrt{\mu}} \le \sqrt{\beta\epsilon} \tag{28}$$

for every $i = 1, 2, \ldots, n$. These properties are valuable in keeping track of the solution convergence; they induce termination when the relative error in the complementarity is small.

Below, we extend the above approach for solving LCP(q, M) and NCP(F) to the solution of the NLP problem (1). Towards this purpose, we reformulate the KKT equations (5) as follows:

$$\Psi_\mu(y) := \begin{pmatrix} \psi_\mu(s, \pi) \\ \psi_\mu(z, \hat{n}) \\ s + z - \overline{h} + \underline{h} \\ h(x) + z - \overline{h} \\ \nabla_x f(x) - G(x)\lambda + H(x)v \\ -g(x) \end{pmatrix} = 0. \tag{29}$$

Similar to IP algorithms, a damped Newton-type method is used to solve (29). The nonzero pattern of the new Newton system is the same as that of (6). The way to compute the elements of the diagonal matrices is the only change. For instance, the diagonal matrix that takes the place of S is computed from

$$\nabla_s\psi_\mu(s, \pi) := \text{diag}\left(1 - \frac{s_i}{\sqrt{s_i^2 + \pi_i^2 + 2\mu^k}}\right). \tag{30}$$

The property (26) assures that the new diagonal matrices, like their counterparts in the IP algorithm, are positive definite.

New estimates for the variables are computed from

$$y^{k+1} := y^k + \alpha^k \Delta y, \tag{31}$$

where Δy is the search direction

$$\Delta y = -[\nabla \Psi_\mu(y^k)]^{-1} \Psi_\mu(y^k), \tag{32}$$

and α^k is the *step length* along Δy. A suitable step length α^k should be computed by performing a *line search* along the direction Δy, aiming at achieving a "sufficient" decrease in a *merit function*. Since our intent is to find a solution of $\Psi_\mu(y) = 0$, an obvious merit function would be

$$\Phi_\mu(y) := \frac{1}{2} \Psi_\mu(y)^T \Psi_\mu(y), \tag{33}$$

which is known as the *natural merit function*. When solving the nonlinear system (29), we indeed turn our attention to an unconstrained minimization

$$\min_{y \in \mathbb{R}^q} \Phi_\mu(y). \tag{34}$$

While every solution to (29) is a solution to (34), there may be local minimizers of (34) that are not solutions to (29). Yet, a solution of $\Psi_\mu(y) = 0$ is not necessarily a minimizer of (1) (maximizers and saddle points also satisfy this condition). A *non-monotone* Armijo-type line search, commonly used in the solution of complementarity problems, is as follows [14,11]:

- Given $\sigma_1 \in (0,1]$ and $\alpha_1 \in (0,1)$, find the smallest $m_k \in \{0,1,2,\dots\}$ that satisfies

$$\Phi_\mu(y^k + \alpha_1^{m_k} \Delta y) \le \max_{j=k-l+1:\ k} \Phi_\mu(y^j) - \sigma_1 \alpha_1^{m_k} \Phi_\mu(y^k). \tag{35}$$

- Then, let $\alpha^k = \alpha_1^{m_k}$.

We use $\sigma_1 = 10^{-4}$, $\alpha_1 = 0.75$, and $l = \min\{5, k\}$.

We reduce μ^k in a way quite similar to that used in the IP method. First, we compute the measure

$$\varrho^k := |s^k|^T |\pi^k| + |z^k|^T |\hat{v}^k|, \tag{36}$$

which "mimics" the complementarity gap ρ^k in the IP method. If the iterates converge to an optimum, then $\{\varrho^k\} \to 0$. Thus, we compute μ^{k+1} from

$$\mu^{k+1} := \min\left\{\sigma \frac{\varrho^k}{2p}, 0.9\mu^k\right\}, \tag{37}$$

where σ is as defined in the IP algorithm.

5   Some Preliminary Results

Initially, we developed prototype codes under MATLAB (The MathWorks, Inc.) and tested on the well known IEEE test systems with 30, 57, 118 and 300 buses.

To be able to solve larger problems, and compare CPU times, we also coded all algorithms in Fortran 77 (in double precision arithmetic) and compiled with the -O2 option. The results presented have been produced on a Pentium Pro 200 MHz with 64 MBytes of RAM, running the Linux operating system.

The developed codes been tested on a set of seven power systems that range in size from 30 to 2098 nodes. Some statistics for the test systems are displayed in Table 1, where we give for each power system the total number of nodes ($|\mathcal{N}|$), the number of *generators* ($|\mathcal{G}|$), the number of nodes *eligible* for shunt var control ($|\mathcal{E}|$), the total number of *branches* ($|\mathcal{B}|$), and the number of *transformers* with LTC device ($|\mathcal{T}|$). Also displayed in Table 1 are the number of primal variables (n), the number of equality constraints (m), and the number of inequality constraints subject to lower and upper bounds (p). The problem objective function (operation performance index) employed is the minimization of active-power losses in the transmission system [5].

Each problem has been solved by three different codes, which implement the plain primal-dual IP method that is described in Sect. 2 (code PDM), the predictor-corrector IP method that is described in Sect. 3 (code PCM), and the NIP method that is described in Sect. 4 (code NIP). The numerical results are displayed in Table 2, where for each test run we give the number of iterations to convergence (iters), the CPU times in *seconds* (time), and the final active-power losses (P_{loss}) in percentage of the total active-power consumption. Additionally, for the NIP code, we give the total number of merit function evaluations (Φ-eval). In all instances, the IP and NIP algorithms performed extremely well, with the number of iterations insensitive to the dimension of the problems.

Table 1. Sizes of the test systems and related NLP problems

Problem	$|\mathcal{N}|$	$|\mathcal{G}|$	$|\mathcal{E}|$	$|\mathcal{B}|$	$|\mathcal{T}|$	n	m	p
PS-1	30	6	4	41	4	63	49	44
PS-2	57	7	5	80	15	128	101	84
PS-3	118	54	12	186	9	244	169	193
PS-4	300	69	12	411	50	649	518	431
PS-5	256	68	23	376	50	561	430	387
PS-6	555	126	46	787	85	1194	937	812
PS-7	2098	169	426	3283	239	4434	3600	2932

For the simulations with the NIP algorithm, we use the same initialization of the IP algorithm, as described in [5], the non-monotone line search test to obtain the step length, and the proposed scheme to update the continuation parameter. It should be noted that while the merit function used for the NIP method only guarantees convergence to a stationary point of the Lagrangian function, not necessarily a local minimizer of the NLP problem, in all test runs

Table 2. Computational performance of the IP and NIP algorithms

Problem	PDM code iters	PDM code time	PCM code iters	PCM code time	NIP code iters	NIP code Φ-eval	NIP code time	P_{loss} (%)
PS-1	14	1.82	9	1.16	9	9	1.18	7.65
PS-2	17	2.56	10	1.57	9	9	1.34	13.70
PS-3	18	3.61	11	2.30	10	10	2.01	8.59
PS-4	19	6.84	11	4.26	11	11	4.06	8.15
PS-5	19	5.81	11	3.48	13	13	4.05	5.68
PS-6	20	12.78	13	8.70	12	12	7.80	6.47
PS-7	25	58.71	25	61.77	13	23	31.75	5.42
Total	132	92.13	90	83.24	77	77	52.19	

the documented minimizer was obtained. Notice from Table 2 that only one merit function evaluation per iteration has been performed.

In Table 3, we display the convergence process for the solution of problem PS-3 by a prototype implementation of the NIP algorithm under MATLAB. In such an implementation, we have used a fixed step length $\alpha^k = 0.95$ and chosen $\mu^0 = 0.01$, with the rather simple updating rule $\mu^{k+1} = \max\{0.1\mu^k, 10^{-10}\}$. We have chosen x^0 as given by a load flow solution, with $s^0 = h(x^0) - \underline{h}$, $z^0 = \overline{h} - h(x^0)$, $\pi^0 = e$ and $v^0 = 0$. This implementation in MATLAB has performed extremely well, as we have applied it to the smaller problems PS-1, PS-2, PS-3 and PS-4.

Table 3. Convergence process for problem PS-3 solved by code NIP

k	ν_1^k	ν_2^k	ν_3^k	ν_4^k	$\Psi_\mu(y^k)$
0	4.494×10^{-1}	2.218×10^{-1}	$1.000 \times 10^{+0}$	$1.084 \times 10^{+0}$	$1.499 \times 10^{+1}$
1	2.840×10^{-2}	4.304×10^{-2}	$5.808 \times 10^{+0}$	$3.108 \times 10^{+1}$	$1.726 \times 10^{+4}$
2	3.672×10^{-2}	8.817×10^{-2}	3.793×10^{-1}	$2.534 \times 10^{+0}$	$5.196 \times 10^{+1}$
3	8.029×10^{-2}	1.766×10^{-2}	8.928×10^{-2}	4.672×10^{-1}	9.464×10^{-1}
4	1.034×10^{-1}	8.768×10^{-3}	9.949×10^{-2}	8.750×10^{-2}	2.268×10^{-1}
5	2.821×10^{-1}	7.703×10^{-4}	2.875×10^{-1}	1.371×10^{-2}	9.424×10^{-1}
6	1.732×10^{-1}	1.699×10^{-4}	1.737×10^{-1}	4.283×10^{-3}	3.685×10^{-1}
7	1.298×10^{-2}	1.145×10^{-4}	2.288×10^{-2}	7.737×10^{-4}	3.787×10^{-3}
8	7.688×10^{-4}	1.117×10^{-5}	4.717×10^{-3}	1.505×10^{-4}	1.102×10^{-4}
9	3.563×10^{-2}	5.085×10^{-6}	3.567×10^{-2}	2.054×10^{-5}	5.085×10^{-3}
10	1.825×10^{-3}	1.459×10^{-6}	1.832×10^{-3}	4.307×10^{-6}	1.292×10^{-5}
11	1.370×10^{-4}	9.536×10^{-7}	1.373×10^{-4}	8.556×10^{-7}	4.400×10^{-8}
12	2.808×10^{-5}	5.131×10^{-8}	2.809×10^{-5}	3.993×10^{-8}	1.253×10^{-9}

6   Conclusions

The paper has described how the nonlinear OPF problem can be efficiently solved by IP and NIP methods for nonlinear programming. Ongoing research to further improve the IP algorithm performance involves the computation of α and reduction of μ based on merit functions, as well as more elaborate starting point choices. Given the ability of the NIP algorithm to start from arbitrary points, and the observation that the IP and NIP algorithms basically use the same system matrix, we are currently working on an improved OPF algorithm that combines together the IP and NIP approaches. Switching from one approach to the other demands no changes of the linear algebra kernel. Applications of other NCP-functions are being studied as well.

Acknowledgments

The authors thank the financial support from CAPES and UFPE in Brazil, and NSERC in Canada.

References

1. V. H. Quintana, G. L. Torres, and J. Medina-Palomo, "Interior-point methods and their applications to power systems: A classification of publications and software codes," To appear in *IEEE Trans. on Power Systems*, Paper No. 98 SM 244.

2. K. A. Clements, P. W. Davis, and K. D. Frey, "Treatment of inequality constraints in power system state estimation," *IEEE Trans. on Power Systems*, vol. 10, pp. 567–573, May 1995.

3. Y. Wu, A. S. Debs, and R. E. Marsten, "A direct nonlinear predictor-corrector primal-dual interior point algorithm for optimal power flow," *IEEE Trans. on Power Systems*, vol. 9, pp. 876–883, May 1994.

4. S. Granville, "Optimal reactive dispatch through interior point methods," *IEEE Trans. on Power Systems*, vol. 9, pp. 136–146, Feb. 1994.

5. G. L. Torres and V. H. Quintana, "An interior point method for nonlinear optimal power flow using voltage rectangular coordinates," To appear in *IEEE Trans. on Power Systems*, Paper no. PE-010-PWRS-0-12-1997.

6. S. Granville, J. C. O. Mello, and A. C. G. Melo, "Application of interior point methods to power flow unsolvability," *IEEE Trans. on Power Systems*, vol. 11, pp. 1096–1103, May 1996.

7. G. D. Irisarri, X. Wang, J. Tong, and S. Mokhtari, "Maximum loadability of power systems using interior point non-linear optimization method," *IEEE Trans. on Power Systems*, vol. 12, pp. 162–172, Feb. 1997.

8. I. J. Lustig, R. E. Marsten, and D. F. Shanno, "Computational experience with a primal-dual interior point method for linear programming," *Linear Algebra and Its Applications*, vol. 152, pp. 191–222, 1991.

9. A. V. Fiacco and G. P. McCormick, *Nonlinear Programming: Sequential Unconstrained Minimization Techniques*. John Wiley & Sons, 1968.

10. S. Mehrotra, "On the implementation of a primal-dual interior point method," *SIAM Journal on Optimization*, vol. 2, pp. 575–601, 1992.

11. B. Chen and P. T. Harker, "A non-interior-point continuation method for linear complementarity problems," *SIAM Journal on Matrix Analysis and Applications*, vol. 14, pp. 1168–1190, Oct. 1993.

12. C. Kanzow, "Some noninterior continuation methods for linear complementarity problems," *SIAM Journal on Matrix Analysis and Applications*, vol. 17, pp. 851–868, Oct. 1996.

13. C. Kanzow, "Some equation-based methods for the nonlinear complementarity problem," *Optimization Methods and Software*, vol. 3, pp. 327–340, 1994.

14. J. V. Burke and S. Xu, "The global linear convergence of a non-interior path-following algorithm for linear complementarity problems," Technical Report, Department of Mathematics, University of Washington, Seattle, WA, Sept. 1996.

Decentralized Control of Uncertain Systems via Sensitivity Models

Dave A. Schoenwald[1] and Ümit Özgüner[2]

[1] Instrumentation and Controls Division, Oak Ridge National Laboratory,
   P. O. Box 2008, Oak Ridge, TN 37831-6003 USA
   Email: schoenwaldda@ornl.gov
[2] Department of Electrical Engineering, The Ohio State University, 2015 Neil
   Avenue, Columbus, OH 43210-1272 USA
   Email: umit@eng.ohio-state.edu

Abstract. In this paper, we present a decentralized strategy for optimal control of interconnected systems exhibiting parametric uncertainty. First, we demonstrate that sensitivity models for linear interconnected systems can be generated at each subsystem using only locally available information. This is generalized to specific classes of nonlinear systems as well. Second, we present an optimal control law that incorporates sensitivity functions in the feedback path. The control scheme is completely decentralized and is proposed as a means of making the closed loop system less sensitive to parameter deviations. Finally, we give an example of an interconnected system and show how this control strategy is implemented. The results are compared to a decentralized optimal control law that does not utilize sensitivity functions.

1 Introduction

The use of sensitivity functions in control theory to make the closed loop system less susceptible to changes in plant parameters has been studied for several decades (see survey by Kokotović and Rutman for an early history [6]). In particular, methods of generating sensitivity functions such that they can be utilized on-line in a control system have been extensively researched. Sensitivity models are a means of generating these sensitivity functions from the nominal plant model. However, very little effort has been spent in generating sensitivity models for coupled subsystems in a decentralized manner.

But, as an additional tool for decentralized control, it is of interest to determine the feasibility of generating these models using only local signals for interconnected systems. That is, we wish to investigate the possibility that the i-th subsystem's output sensitivity function can be generated using only plant signals from the i-th subsystem. This is important if one wants to use sensitivity functions in a decentralized control environment. For instance, one could use decentralized sensitivity functions for self-tuning control (tuning the gains of a control law when some of the plant parameters are unknown) as is done in the work of Hung [4].

In this paper, decentralized sensitivity models are suggested for certain classes of linear systems. It is shown that decentralized sensitivity models are possible for nonlinear systems as well under some assumptions. Seminal work on decentralized control of nonlinear systems appears in [1]. Ultimately, the importance of this paper is to show that any control law that utilizes sensitivity functions (e.g. adaptive or optimal control laws) can be done in a decentralized framework even for nonlinear systems. Thus, the results of this paper could be very practical provided one wishes to employ control laws that make use of sensitivity functions.

The paper is organized as follows. The next section demonstrates the fact that sensitivity models can be decentralized for coupled systems. Section 3 details a decentralized optimal control strategy that utilizes locally generated sensitivity models to minimize the effects of parameter variations on closed-loop performance. Section 4 presents an example consisting of two coupled inverted penduli with unknown natural frequencies of oscillation. The optimal control strategy outlined in this paper is compared to one in which no sensitivity model is utilized for this example. Finally, Section 5 gives some concluding remarks.

2 Decentralized Sensitivity Models

Consider Fig. 1 which depicts two interconnected MIMO linear systems. The outputs, Y_i, are of dimension p_i, $i = 1, 2$. The inputs, U_i, are of dimension m_i. The unknown parameter vectors, α_i, β_i, γ_i, are of dimensions n_i, r_i, and q_i, respectively. The transfer matrices, Q_i, W_i, W_{ij}, are of compatible dimensions. All vectors and matrices are functions of the Laplace Transform complex variable, s. The transfer matrices, Q_i, represent dynamic feedback from the outputs to the inputs. The transfer matrices, W_{ij}, represent coupling terms between the subsystems. The transfer matrices, W_i, represent the primary dynamics between plant inputs U_i and plant outputs Y_i.

The block transfer matrix of the entire system can be obtained by writing input-output relationships as follows

$$Y_1 = W_1(U_1 - Q_1Y_1) + W_{21}(U_2 - Q_2Y_2) \tag{1}$$
$$Y_2 = W_2(U_2 - Q_2Y_2) + W_{12}(U_1 - Q_1Y_1) . \tag{2}$$

Letting

$$\Delta_1 = I + W_1Q_1 - W_{21}Q_2[I + W_2Q_2]^{-1}W_{12}Q_1 \tag{3}$$
$$\Delta_2 = I + W_2Q_2 - W_{12}Q_1[I + W_1Q_1]^{-1}W_{21}Q_2 \tag{4}$$

where I is the identity matrix, we obtain the input-output description of the system (after some algebraic manipulation)

$$\begin{bmatrix} Y_1 \\ Y_2 \end{bmatrix} = \begin{bmatrix} F_{11} & F_{12} \\ F_{21} & F_{22} \end{bmatrix} \begin{bmatrix} U_1 \\ U_2 \end{bmatrix} \tag{5}$$

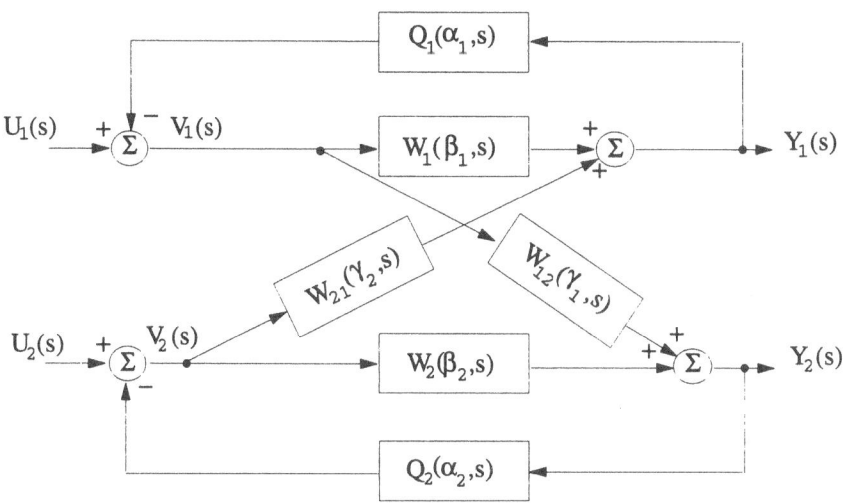

Fig. 1. MIMO system with coupled linear subsystems.

with

$$F_{11} = \Delta_1^{-1}[W_1 - W_{21}Q_2(I + W_2Q_2)^{-1}W_{12}] \tag{6}$$
$$F_{12} = \Delta_1^{-1}[W_{21} - W_{21}Q_2(I + W_2Q_2)^{-1}W_2] \tag{7}$$
$$F_{21} = \Delta_2^{-1}[W_{12} - W_{12}Q_1(I + W_1Q_1)^{-1}W_1] \tag{8}$$
$$F_{22} = \Delta_2^{-1}[W_2 - W_{12}Q_1(I + W_1Q_1)^{-1}W_{21}]. \tag{9}$$

We are interested in output sensitivity vectors of the system with respect to the unknown parameter vectors, α_i, β_i, and γ_i. It is sufficient to study sensitivity vectors of the first subsystem since the two subsystems are symmetric as is easily observed.

We proceed with the sensitivity of Y_1 with respect to α_1 which begins by utilizing (1)

$$\frac{\partial Y_1}{\partial \alpha_{1i}} = \frac{\partial F_{11}}{\partial \alpha_{1i}}U_1 + \frac{\partial F_{12}}{\partial \alpha_{1i}}U_2 \tag{10}$$

where α_{1i} is the i-th component of the unknown parameter vector. It is important to note here that the commutativity of SISO systems that yielded the Wilkie-Perkins result [9] of just one sensitivity model (plus the plant itself) to simultaneously generate all the sensitivity functions does not apply. This means that the process of applying $Y_1(s)$ as an input to a sensitivity filter-sensitivity model pair must be executed n_1 times (for the α_1 parameter vector). Thus, the output sensitivity vectors of this paper will be with respect to the i-th parameter of the vector in question. This in no way affects the versatility of the decentralized models in this section since it would apply to SISO systems as well.

Continuing with the analysis, the sensitivity of F_{11} with respect to the parameter α_{1i} is

$$\frac{\partial F_{11}}{\partial \alpha_{1i}} = -\Delta_1^{-1} \frac{\partial \Delta_1}{\partial \alpha_{1i}} \Delta_1^{-1} [W_1 - W_{21} Q_2 (I + W_2 Q_2)^{-1} W_{12}] \qquad (11)$$

where

$$\frac{\partial \Delta_1}{\partial \alpha_{1i}} = [W_1 - W_{21} Q_2 (I + W_2 Q_2)^{-1} W_{12}] \frac{\partial Q_1}{\partial \alpha_{1i}} . \qquad (12)$$

With this equation, we can rewrite $\frac{\partial F_{11}}{\partial \alpha_{1i}}$ as (after some algebraic manipulation)

$$\frac{\partial F_{11}}{\partial \alpha_{1i}} = -F_{11} \frac{\partial Q_1}{\partial \alpha_{1i}} F_{11} . \qquad (13)$$

Likewise, via similar steps, one can obtain

$$\frac{\partial F_{12}}{\partial \alpha_{1i}} = -F_{11} \frac{\partial Q_1}{\partial \alpha_{1i}} F_{12} . \qquad (14)$$

This leads to

$$\frac{\partial Y_1}{\partial \alpha_{1i}} = -F_{11} \frac{\partial Q_1}{\partial \alpha_{1i}} (F_{11} U_1 + F_{12} U_2) . \qquad (15)$$

Noting that the term in parentheses is merely (1), we have the result

$$\frac{\partial Y_1}{\partial \alpha_{1i}} = -F_{11} \frac{\partial Q_1}{\partial \alpha_{1i}} Y_1 \qquad (16)$$

which is a completely decentralized result. The output sensitivity vector depends only on the output signal itself plus some transfer matrices.

Several comments are in order at this juncture. First, one would normally be interested in the sensitivity functions evaluated about some known nominal parameter values. In this case, the above partial derivatives are all evaluated about these nominal values. This dependence on the nominal values is suppressed in the analysis to make notation simpler, but it should be assumed. Second, the above result and the ones to follow are decentralized in the sense that only local (i.e. within the given subsystem) signals are needed to compute the sensitivity functions. This is generally what one means by decentralized. The fact that transfer matrices from other subsystems are needed to compute these sensitivity functions does not detract from the decentralized nature of these results. It is generally assumed in the decentralized control literature that models can be exchanged across subsystems even though signals are not. This is because in a real-time environment, the nominal models will have been known for some time whereas the signals are being generated at the moment (though there are some results that do use only local models).

Proceeding in a likewise manner for the output sensitivity vector with respect to β_1,

$$\frac{\partial Y_1}{\partial \beta_{1i}} = \frac{\partial F_{11}}{\partial \beta_{1i}} U_1 + \frac{\partial F_{12}}{\partial \beta_{1i}} U_2 \tag{17}$$

where (after some manipulation)

$$\frac{\partial F_{11}}{\partial \beta_{1i}} = \Delta_1^{-1} \frac{\partial W_1}{\partial \beta_{1i}} [-Q_1 F_{11} + I] \tag{18}$$

$$\frac{\partial F_{12}}{\partial \beta_{1i}} = \Delta_1^{-1} \frac{\partial W_1}{\partial \beta_{1i}} Q_1 F_{12} . \tag{19}$$

Finally, we obtain

$$\frac{\partial Y_1}{\partial \beta_{1i}} = \Delta_1^{-1} \frac{\partial W_1}{\partial \beta_{1i}} V_1 \tag{20}$$

where

$$V_1 = U_1 - Q_1 Y_1 \tag{21}$$

and represents the error signal between the reference inputs and the feedback signal. Thus, this sensitivity vector can also be generated via a local sensitivity model.

The sensitivity of Y_1 with respect to γ_{1i} is also of interest since one might want to know how the output of the first subsystem is affected by parametric uncertainty in the cross-coupling from subsystem 1 to subsystem 2. We start as before with

$$\frac{\partial Y_1}{\partial \gamma_{1i}} = \frac{\partial F_{11}}{\partial \gamma_{1i}} U_1 + \frac{\partial F_{12}}{\partial \gamma_{1i}} U_2 \tag{22}$$

where

$$\frac{\partial F_{11}}{\partial \gamma_{1i}} = \Delta_1^{-1} W_{21} Q_2 (I + W_2 Q_2)^{-1} \frac{\partial W_{12}}{\partial \gamma_{1i}} [Q_1 F_{12} - I] \tag{23}$$

$$\frac{\partial F_{12}}{\partial \gamma_{1i}} = \Delta_1^{-1} W_{21} Q_2 (I + W_2 Q_2)^{-1} \frac{\partial W_{12}}{\partial \gamma_{1i}} Q_1 F_{12} . \tag{24}$$

Combining these equations with (22) yields

$$\frac{\partial Y_1}{\partial \gamma_{1i}} = \Delta_1^{-1} W_{21} Q_2 (I + W_2 Q_2)^{-1} \frac{\partial W_{12}}{\partial \gamma_{1i}} [Q_1 (F_{12} - F_{11}) U_1 - V_1] \tag{25}$$

which again shows that it can be generated in a decentralized manner.

The three remaining output sensitivity vectors of the first subsystem do not possess the same decentralized features as the above sensitivity vectors. However, they are of interest to us since they still maintain some elements of a decentralized structure. Since the procedure for generating these sensitivity vectors has been established, the results are simply stated as follows

$$\frac{\partial Y_1}{\partial \alpha_{2i}} = -F_{12} \frac{\partial Q_2}{\partial \alpha_{2i}} Y_2 \tag{26}$$

$$\frac{\partial Y_1}{\partial \beta_{2i}} = -\Delta_1^{-1} W_{21} Q_2 (I + W_2 Q_2)^{-1} \frac{\partial W_2}{\partial \beta_{2i}} V_2 \tag{27}$$

$$\frac{\partial Y_1}{\partial \gamma_{2i}} = \Delta_1^{-1} \frac{\partial W_{21}}{\partial \gamma_{2i}} V_2 . \tag{28}$$

From above, one can see that each of these sensitivity vectors requires signals from only the second subsystem. In this sense, they can be generated by decentralized sensitivity models, but they would only be available at the second subsystem. Presumably, one would want these sensitivity vectors at the first subsystem since they do reflect Y_1's dependence on unknown parameters in the second subsystem. However, in some cases one may be able to use this information at the second subsystem where it can be generated via local sensitivity models. This might be the case in adaptive control where the sensitivity of the other subsystem's output to local parameters is needed to update a decentralized control law.

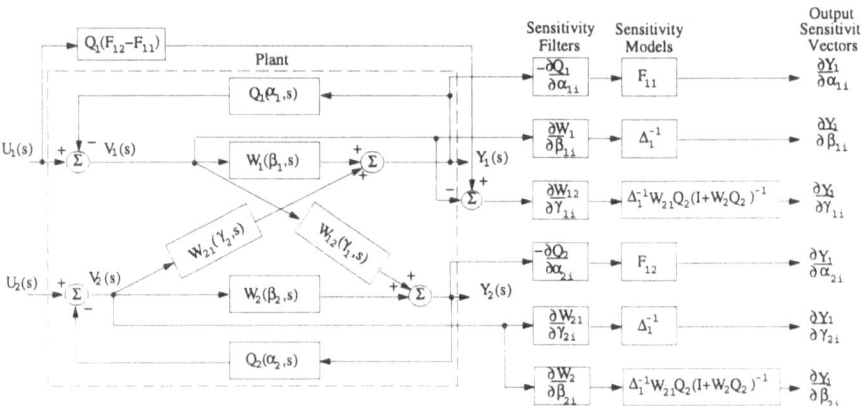

Fig. 2. Sensitivity model configurations for linear coupled subsystems.

Fig. 2 shows the configuration for generating the sensitivities of the first subsystem for the plant model in Fig. 1. Some of the sensitivity models can serve for two different sensitivity vectors, but each of the sensitivity vectors requires a different sensitivity filter. One drawback of decentralized sensitivity models over the centralized case is that the sensitivity models are no longer just a copy of the plant model. As is apparent in Fig. 2, the sensitivity models cannot be written as just a plant model, but some do have physical meaning. For instance, the sensitivity models for $\frac{\partial Y_1}{\partial \alpha_{1i}}$ and $\frac{\partial Y_1}{\partial \alpha_{2i}}$ which are F_{11} and F_{12}, respectively, represent the plant evaluated at the nominal parameter values with $U_2 = 0$ and $U_1 = 0$, respectively, and the other input coming from its respective sensitivity filter. The other sensitivity models are more complicated and are not able to be generated from a copy of the plant model in any obvious way.

As mentioned earlier, the sensitivities of Y_2 with respect to all of the unknown parameter vectors can be generated by exactly the dual of the above sensitivity models. With this figure one can utilize sensitivity-based control designs in a decentralized framework. Thus, well-known sensitivity control schemes (see [2] for examples) can be utilized at each subsystem. In particular, gradient adaptive control techniques utilizing sensitivity functions would enable one to update subsystem parameters in a completely decentralized setting.

The next logical step in generating decentralized sensitivity models is to look at the nonlinear case. It is much more difficult to produce local sensitivity models for coupled nonlinear systems because first-order partials of vector fields will still depend on the state. This implies that partials of coupling terms will require the use of the state vector from another subsystem which cannot be measured or generated locally. This motivates the analysis of less general nonlinear systems where perhaps only one subsystem needs access to a sensitivity model. Consider the following two-channel system depicted in Fig. 3

$$\dot{x}_1 = f_1(x_1, \alpha_1) + B_1 u_1 + A_{12} x_2, \quad y_1 = h_1(x_1, \alpha_1) \tag{29}$$
$$\dot{x}_2 = A_2(\alpha_2) x_2 + B_2 u_2 + A_{21} x_1, \quad y_2 = C_2(\alpha_2) x_2$$

where only f_1 and h_1 are assumed to be nonlinear. These vector fields are also assumed to be smooth in their arguments, x_1 and α_1. The second subsystem is completely linear and the coupling terms are linear as well. The terms B_i and A_{ij} are assumed to be independent of α_i.

To generate the sensitivity model for y_1 with respect to α_1, we need the sensitivity models for both x_1 and x_2 with respect to α_1 due to the cross-coupling. Differentiating, we obtain

$$\frac{\partial \dot{x}_1}{\partial \alpha_1} = \frac{\partial f_1}{\partial x_1} \frac{\partial x_1}{\partial \alpha_1} + A_{12} \frac{\partial x_2}{\partial \alpha_1} + \frac{\partial f_1}{\partial \alpha_1}$$
$$\frac{\partial y_1}{\partial \alpha_1} = \frac{\partial h_1}{\partial x_1} \frac{\partial x_1}{\partial \alpha_1} + \frac{\partial h_1}{\partial \alpha_1} \tag{30}$$
$$\frac{\partial \dot{x}_2}{\partial \alpha_1} = A_2(\alpha_2) \frac{\partial x_2}{\partial \alpha_1} + A_{21} \frac{\partial x_1}{\partial \alpha_1}$$
$$\frac{\partial y_2}{\partial \alpha_1} = C_2(\alpha_2) \frac{\partial x_2}{\partial \alpha_1}$$

where it is noted that u_i is assumed independent of α_i.

From the model (30), it can be seen that the output and state sensitivities can all be generated in a decentralized framework. This is because state/output sensitivities form a system with a model requiring only matrices and signals that can be computed locally. The cross sensitivity $\frac{\partial x_2}{\partial \alpha_1}$ can be calculated at

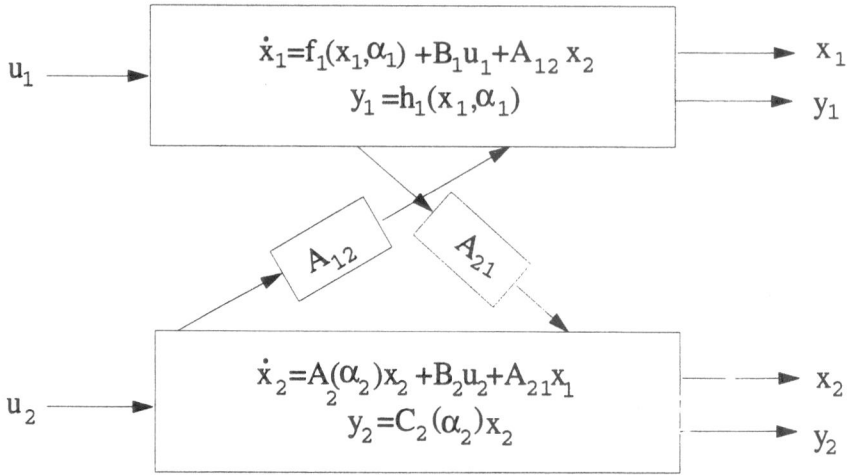

Fig. 3. Two-channel coupled nonlinear system.

subsystem 1 since the above model shows that it too does not require any signals from subsystem 2. The same will not hold true for the sensitivities with respect to α_2 at the second subsystem since they will depend on $\frac{\partial f_1}{\partial x_1}$ which in general is a nonlinear function of x_1. Thus, the state x_1 would be needed at subsystem 2 to compute the sensitivities with respect to α_2 violating the decentralization constraint. Fig. 4 illustrates the manner in which the sensitivity functions of the system in (29) are computed.

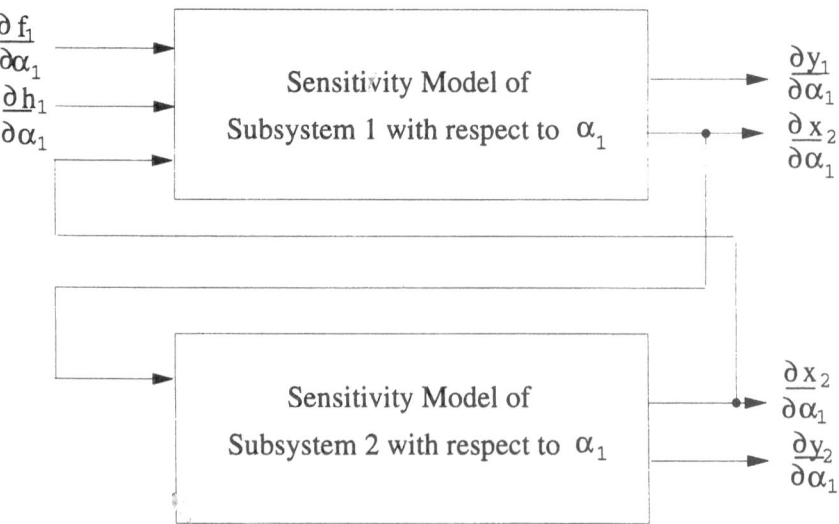

Fig. 4. Sensitivity models for coupled nonlinear systems.

If the control vector fields B_i are not constant (e.g. $g_i(x_i)$ or $g_i(x_i, \alpha_i)$) then the terms $\frac{\partial g_i}{\partial \alpha_i}$ will be multiplying u_i making it impossible for the cross sensitivities to be generated locally (i.e. u_j would be needed in the ith subsystem's sensitivity model). Note also that the sensitivity model (30) is a linear system dependent on Jacobians of the $f_{i(j)}$ vectors with respect to α_i. The model is, however, time-varying since the Jacobians will depend on x_i in general. But the model will not depend on the unknown parameters since all parameters will be evaluated at their nominal values.

3   Decentralized Optimal Control via Sensitivity Functions

It is now of interest to apply the above decentralized sensitivity models to performance issues associated with decentralized control. In particular, we concern ourselves with locally optimal control laws which include sensitivity functions in the feedback loop as a means of desensitizing the system from parameter variations. The framework pursued here is that of linear subsystems with linear couplings similar to that of [8] but utilizing sensitivity functions to address parametric uncertainty in an optimal manner. A version of the centralized case appears in [2]. The performance index consists of a sum of N local cost criterions where N is the number of subsystems.

Formally, we have

$$\dot{x}_i = A_i(\alpha_i)x_i + \sum_{\substack{j=1 \\ j \neq i}}^{N} A_{ij}x_j + B_i u_i, \quad i = 1, \ldots, N \tag{31}$$

with $x_i \in R^{n_i}$, $u_i \in R^{m_i}$, and $\alpha_i \in R_{p_i}$ where α_i are the vectors of unknown parameters. It is desired to minimize the quadratic cost criterion

$$J = \sum_{i=1}^{N} \int_0^\infty \left(x_i^T Q_i x_i + u_i^T R_i u_i + \sum_{j=1}^{N} \lambda_{ij}^T S_{ij} \lambda_{ij} \right) dt \tag{32}$$

via local state feedback where Q_i, S_{ij} are positive semidefinite matrices, R_i are positive definite matrices, and $\lambda_{ij} = \frac{\partial x_j}{\partial \alpha_i} \big|_{\alpha_i = \alpha_i^n}$ are the sensitivity vectors associated with the i-th subsystem with α_i^n the nominal parameter values. As was demonstrated in the previous section, the cross-sensitivity functions $\lambda_{ij}, j \neq i$ can be generated at the i-th subsystem with only local information needed.

The sensitivity models are described by the following linear time-invariant differential equations

$$\dot{\lambda}_{ii} = \frac{\partial A_i}{\partial \alpha_i} x_i + A_i \lambda_{ii} + \sum_{\substack{j=1 \\ j \neq i}}^{N} A_{ij} \lambda_{ij} \tag{33}$$

$$\dot{\lambda}_{ij} = A_j \lambda_{ij} + \sum_{\substack{k=1 \\ k \neq j}} A_{jk} \lambda_{ik} \tag{34}$$

which again demonstrate the decentralized nature of sensitivity models for the system (31). Define the augmented state vector

$$\tilde{x}_i = \left[x_i^T \; \lambda_{ii}^T \; \lambda_{ij}^T \right]^T , \tag{35}$$

and one obtains the following linear time-invariant system

$$\begin{bmatrix} \dot{x}_i \\ \dot{\lambda}_{ii} \\ \dot{\lambda}_{ij} \end{bmatrix} = \begin{bmatrix} A_i \mid_{\alpha_i^n} & 0 & 0 \\ \frac{\partial A_i}{\partial \alpha_i} \mid_{\alpha_i^n} & A_i \mid_{\alpha_i^n} & \sum_{j=1}^{N} A_{ij} \\ 0 & A_{ji} & A_j \mid_{\alpha_j^n} + \sum_{\substack{k=1 \\ k \neq i,j}}^{N} A_{jk} \end{bmatrix} \begin{bmatrix} x_i \\ \lambda_{ii} \\ \lambda_{ij} \end{bmatrix} \tag{36}$$

$$+ \sum_{\substack{j=1 \\ j \neq i}}^{N} \begin{bmatrix} A_{ij} \; 0 \; 0 \\ 0 \; \; 0 \; 0 \\ 0 \; \; 0 \; 0 \end{bmatrix} \begin{bmatrix} x_j \\ \lambda_{jj} \\ \lambda_{ji} \end{bmatrix} + \begin{bmatrix} B_i \\ 0 \\ 0 \end{bmatrix} u_i$$

where the outputs are defined accordingly but not needed in this analysis. The compact form

$$\dot{\tilde{x}}_i = \tilde{A}_i \tilde{x}_i + \sum_{\substack{j=1 \\ j \neq i}}^{N} \tilde{A}_{ij} \tilde{x}_j + \tilde{B}_i u_i \tag{37}$$

which immediately follows from (36) leads to the full augmented state space description

$$\dot{\tilde{x}} = \tilde{A} \tilde{x} + \tilde{A}_c \tilde{x} + \tilde{B} u \tag{38}$$

where $\tilde{x} = \left[\tilde{x}_1^T \cdots \tilde{x}_N^T \right]^T$, $\tilde{A} = \text{block-diag} \left[\tilde{A}_1 \cdots \tilde{A}_N \right]$, $\tilde{A}_c = \left[\tilde{A}_{ij} \right]$, $j \neq i$, $\tilde{B} = \text{block-diag} \left[\tilde{B}_1 \cdots \tilde{B}_N \right]$, and $u = \left[u_1^T \cdots u_N^T \right]$.

The performance criterion (32) can be re-expressed in these new coordinates as

$$J = \sum_{i=1}^{N} \int_0^{\infty} \left(\tilde{x}_i^T \tilde{Q}_i \tilde{x}_i + u_i^T R_i u_i \right) dt \tag{39}$$

where $\tilde{Q}_i = \text{block-diag}\,[Q_i\ S_{ii}\ S_{ij}]$. This is written in the full state space as

$$J = \int_0^\infty \left(\tilde{x}^T \tilde{Q}\tilde{x} + u^T Ru\right) dt \tag{40}$$

where $\tilde{Q} = \text{block-diag}\left[\tilde{Q}_1 \cdots \tilde{Q}_N\right]$ and $R = \text{block-diag}\,[R_1 \cdots R_N]$. Note that $\tilde{Q}$ and R will be positive semidefinite and positive definite, respectively, if Q_i, S_{ij} and R_i are positive semidefinite and positive definite, respectively.

It is assumed that the decoupled subsystems

$$\dot{\tilde{x}} = \tilde{A}\tilde{x} + \tilde{B}u \tag{41}$$

are controllable, i.e., $(\tilde{A}, \tilde{B})$ are a controllable pair. Thus, the cost criterion (40) can be minimized by solving the linear quadratic regulator separately for each subsystem. That is, let

$$u = -K\tilde{x} \tag{42}$$

where $K = \text{block-diag}\,[K_1 \cdots K_N]$. These K_i are computed by solving the algebraic Riccati equation

$$\tilde{A}^T P + P\tilde{A} - P\tilde{B}R^{-1}\tilde{B}^T P + \tilde{Q} = 0 \tag{43}$$

for the unique positive definite matrix $P = \text{block-diag}\,[P_1 \cdots P_N]$. Because of the structure of $\tilde{A}, \tilde{B}, \tilde{Q}, R$, the solution P will automatically be in this block-diagonal form. The feedback K is given as

$$K = R^{-1}\tilde{B}^T P \tag{44}$$

which produces the optimal cost

$$J^* = \tilde{x}^T(0) P\tilde{x}(0) \tag{45}$$

where it is noted that the initial conditions on the sensitivity vectors are always zero.

The control law (42) is completely decentralized, $u = -K_i\tilde{x}_i$, which means that each subsystem can be regulated with only locally available information even in the presence of parametric uncertainty. Indeed, it is the use of locally generated sensitivity functions that sets this strategy apart from others. Fig. 5 illustrates this method. Of course, since the interconnections are ignored in the Riccati equation (43), the control law (42) will generally be suboptimal. However, because each subsystem is autonomously driven, this strategy will be robust to a wide range of uncertainties in the interconnections. Since the closed loop system is given by

$$\dot{\tilde{x}} = (\tilde{A} - \tilde{B}K + \tilde{A}_c)\tilde{x}, \tag{46}$$

the interconnections will behave as regular (as opposed to singular) perturbations. In the event of weak coupling, their impact will be small making this

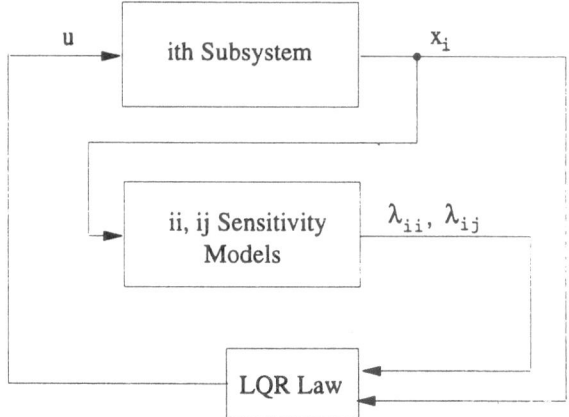

Fig. 5. Decentralized optimal control with sensitivity models.

suboptimal strategy quite effective. The coupling matrix $\tilde{A}_c$ can be taken into account by applying functional minimization schemes (e.g. see [3,5]) that involve coupled Lyapunov equations and iterative procedures. All the necessary terms and definitions for including the interconnections are present in this analysis. This is a straightforward extension of the above method but is omitted here. Finally, if the subsystems are nonlinear then the sensitivity models can be decentralized under the structure in the last section. However, the optimal control strategy pursued would have to involve solving the Hamilton-Jacobi equation unless some linearization procedure were implemented.

4   Example

We consider a system consisting of two inverted penduli coupled by a spring subject to two independent torque inputs as shown in Fig. 6. Physically, this system is analogous to two one-link manipulators joined together by a string, cable, or other spring-like medium. The deflections from vertical are assumed to be small enough such that the gravity term can be linearized. The equations of motion are [8]

$$m\ell_1^2 \ddot{\theta}_1 = mg\ell_1\theta_1 - ka^2(\theta_1 - \theta_2) + u_1 \qquad (47)$$
$$m\ell_2^2 \ddot{\theta}_2 = mg\ell_2\theta_2 - ka^2(\theta_2 - \theta_1) + u_2$$

where all parameters are defined in Fig. 6 except for g which is the gravitational constant. The state vector is chosen as $x_1 = (\theta_1, \dot{\theta}_1)^T$, the input vector is $u = (u_1, u_2)^T$, and the uncertain parameters are $\alpha_i = \frac{g}{\ell_i}$. These parameters physically represent the squares of the natural frequencies of oscillation of the decoupled penduli. We assume some uncertainty from their nominal values.

With this notation the system

$$
\dot{x} =
\begin{bmatrix}
0 & 1 \cdots & 0 & 0 \\
\alpha_1 - \frac{ka^2}{m\ell_1^2} & 0 \cdots & \frac{ka^2}{m\ell_1^2} & 0 \\
\vdots & \vdots \ddots & \vdots & \vdots \\
0 & 0 \cdots & 0 & 1 \\
\frac{ka^2}{m\ell_2^2} & 0 \cdots \alpha_2 - \frac{ka^2}{m\ell_2^2} & 0
\end{bmatrix}
x +
\begin{bmatrix}
0 & \cdots & 0 \\
\frac{1}{m\ell_1^2} & \cdots & 0 \\
\vdots & & \vdots \\
0 & \cdots & 0 \\
0 & \cdots & \frac{1}{m\ell_2^2}
\end{bmatrix}
u
\tag{48}
$$

is written which falls into the format of (31).

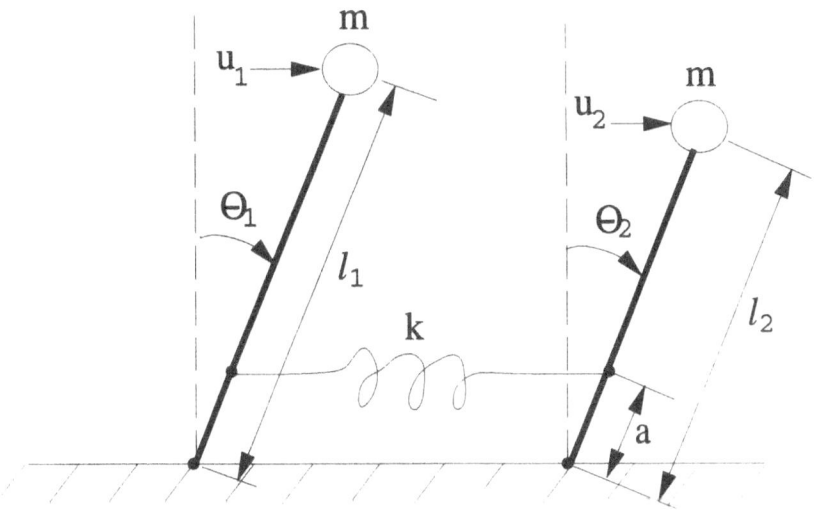

Fig. 6. Inverted penduli coupled by a spring.

The next task is to generate the sensitivity models of the system (48). Utilizing (33)-(34), the sensitivity vectors

$$
\dot{\lambda}_{11} =
\begin{bmatrix} 0 & 0 \\ 1 & 0 \end{bmatrix} x_1 +
\begin{bmatrix} 0 & 1 \\ \alpha_1^n - \frac{ka^2}{m\ell_1^2} & 0 \end{bmatrix} \lambda_{11} +
\begin{bmatrix} 0 & 0 \\ \frac{ka^2}{m\ell_1^2} & 0 \end{bmatrix} \lambda_{12}
$$

$$
\dot{\lambda}_{12} =
\begin{bmatrix} 0 & 1 \\ \alpha_2^n - \frac{ka^2}{m\ell_2^2} & 0 \end{bmatrix} \lambda_{12} +
\begin{bmatrix} 0 & 0 \\ \frac{ka^2}{m\ell_2^2} & 0 \end{bmatrix} \lambda_{11}
\tag{49}
$$

$$
\dot{\lambda}_{21} =
\begin{bmatrix} 0 & 1 \\ \alpha_1^n - \frac{ka^2}{m\ell_1^2} & 0 \end{bmatrix} \lambda_{21} +
\begin{bmatrix} 0 & 0 \\ \frac{ka^2}{m\ell_1^2} & 0 \end{bmatrix} \lambda_{22}
$$

$$
\dot{\lambda}_{22} =
\begin{bmatrix} 0 & 0 \\ 1 & 0 \end{bmatrix} x_2 +
\begin{bmatrix} 0 & 1 \\ \alpha_2^n - \frac{ka^2}{m\ell_2^2} & 0 \end{bmatrix} \lambda_{22} +
\begin{bmatrix} 0 & 0 \\ \frac{ka^2}{m\ell_2^2} & 0 \end{bmatrix} \lambda_{21}
$$

are generated. The augmented state vector $\tilde{x}$ is formed as follows

$$\tilde{x} = \begin{bmatrix} x_1^T & \lambda_{11}^T & \lambda_{12}^T & x_2^T & \lambda_{22}^T & \lambda_{21}^T \end{bmatrix}^T \tag{50}$$

which is 12th order. We Choose $ka^2 = 1\text{N·m}$, $m\ell_1^2 = 1\text{kg·m}^2$, $m\ell_2^2 = 0.5\text{kg·m}^2$, $\alpha_1^n = \frac{g}{\ell_1} = 1\frac{1}{s^2}$, and $\alpha_2^n = \frac{g}{\ell_2} = 2\frac{1}{s^2}$.

The quadratic cost criterion is chosen such that all states and sensitivity functions are weighted equally, i.e., $\tilde{Q}$ is a 12x12 identity matrix. Likewise, R is selected to be a 2x2 identity matrix. Analysis simulated on MATLAB solves the Riccati equation (43) and implements the decentralized control strategy of the last section. For comparison purposes, a decentralized design is carried out on the same system without using sensitivity models. That is, the nominal parameter values were taken as exact in the control design. The uncertainties tested were 10% and 20%, respectively, i.e., the true values were $\alpha_1 = 1.1$, $\alpha_2 = 2.2$, and $\alpha_1 = 1.2$, $\alpha_2 = 2.4$ for the two simulation runs. For both designs, the controllability assumption is satisfied.

Table 1. Closed loop poles of simulation runs with no uncertainty.

	No sensitivity model	Sensitivity model
No interconnections	-0.866±j0.5 -1.414 -1.414	-0.233±j1.255 -0.924±j0.582 -1.376, -1.058 -0.23±j1.206 -1.196±j0.41 -1.79, -1.036
Interconnections	-1.116±j1.207 -2.329 0.0	-2.957±j1.52 -0.53±j1.724 -0.136±j1.259 -0.182±j1.04 -0.22±j0.301 -1.189±j0.003

The results are summarized in Tables 1-3. The first column of numbers of each table represent the closed loop poles for the 4th order decentralized design without sensitivity models. The second column of numbers of each table represent the closed loop poles for the 12th order decentralized design with sensitivity models. The first row of each table corresponds to the case of ignoring the interconnection terms whereas the second row includes these terms in the closed loop analysis. The results show that the lower order design without sensitivity models fails to stabilize the closed loop system once inaccuracies in the parameters are introduced. In fact, even with no uncertainty in the parameters, one of the closed loop poles is at the origin once the coupling matrix is included. As the uncertainty is increased this pole moves further into the right half plane. But for all three cases of uncertainty (0, 10%, 20%), the higher

Table 2. Closed loop poles of simulations runs with 10% uncertainty.

	No sensitivity model	Sensitivity model
No interconnections	0.056 -1.788 0.069 -2.898	-0.249±j1.445 -0.193±j0.836 -2.678, -3.629 -0.301±j1.542 -0.131±j0.904 -1.186, -1.187
Interconnections	-1.03±j0.31 -3.059 0.559	-3.405±j0.995 -0.205±j1.78 -0.168±j1.332 -0.072±j0.978 -0.177±j0.648 -1.188±j0.001

Table 3. Closed loop poles of simulation runs with 20% uncertainty.

	No sensitivity model	Sensitivity model
No interconnections	0.109 -1.84 0.135 -2.963	-0.24±j1.459 -0.173±j0.841 -2.736, -3.707 -0.281±j1.563 -0.113±j0.908 -1.186, -1.187
Interconnections	-3.114 0.601 -0.871 -1.177	-3.45±j0.959 -0.183±j1.784 -0.168±j1.341 -0.063±j0.979 -0.163±j0.669 -1.188±j0.001

order design with sensitivity models maintains closed loop stability. The price paid is a higher order system, but the gain is a significant amount of stability robustness with respect to parametric uncertainty. Of course, it must be noted that the true optimal cost J^* will be infinite for all three cases for the lower order design whereas it remains finite for the higher order design.

Finally, some comments are in order concerning the relationship of the above problem to the question of transient stability in power systems. The above system is mathematically similar (though not exact) to a two-machine swing equation model of a power system. The equations for a single machine are quite similar to that of a simple pendulum as noted in [7]. Adding a second machine results in coupling close to the spring connection in the above example. The forces exerted on the penduli correspond to the electrical power delivered by the machines. Thus, some transient stability results can be studied from this example, however in a realistic setting, constraints would have to be imposed on the inputs to the machines. The constraints would be in the form of limiting values to these inputs so that the controllers would be practical in a power systems setting. With this comment, the above example not only represents two coupled robotic manipulators as mentioned earlier but also the aforementioned power system.

5   Concluding Remarks

In this paper, we have shown that sensitivity models of linear interconnected systems can be generated at each subsystem using only local signals. This has even been shown for special cases of nonlinear systems. The utility of this result is that any control algorithm that calls for the use of sensitivity functions to alleviate the problems of parametric uncertainty can be implemented in a decentralized setting. Thus, adaptive control, optimal control, system identification, etc. that call for the use of sensitivity functions can be done on interconnected systems using only the local state or output.

In particular, a decentralized optimal control strategy was presented that incorporates sensitivity functions in an augmented state vector. A cost criterion that penalizes these sensitivity functions is utilized which makes the closed loop optimal control law less sensitive to parameter deviations at each subsystem. Moreover, the control is completely decentralized requiring only the solution of algebraic Riccati equations for the feedback gain matrices.

This scheme is applied to a system consisting of two inverted penduli coupled by a spring. It is compared to the same decentralized control law without the use of sensitivity models. When the true natural frequencies of oscillation are allowed to deviate from their known nominal values, the scheme with sensitivity models maintains closed loop stability even up to 20% variation in parameters. The strategy without sensitivity models fails to stabilize the true system when the parameters are varied. The price paid is a higher order system, but the robustness to closed loop stability with sensitivity models makes it very useful for uncertain systems under decentralization constraints.

Acknowledgement

The authors wish to acknowledge the support of the AFOSR under contract F49620-89-C-0046. The first author gratefully acknowledges the support of the Ohio Aerospace Institute through a doctoral fellowship.

References

1. E. J. Davison, "The decentralized stabilization and control of a class of unknown nonlinear time-varying systems," *Automatica*, vol. 10, pp. 309–316, 1974.
2. P. M. Frank, *Introduction to System Sensitivity Theory*, New York: Academic Press, 1978.
3. J. C. Geromel and J. Bernussou, "Optimal decentralized control of dynamic systems," *Automatica*, vol. 18, pp. 545–557, 1982.

4. S. T. Hung, "Multi-input/multi-output sensitivity points tuning," M. S. Thesis, University of Illinois, Urbana, IL, May 1985.

5. F. Khorrami, S. Tien, and Ü. Özgüner, "DOLORES: A software package for analysis and design of optimal decentralized control," in *Proceedings of the 40th IEEE National Aerospace & Electronics Conference*, Dayton, OH, 1988.

6. P. V. Kokotović and R. S. Rutman, "Sensitivity of automatic control systems (survey)," *Automation and Remote Control*, vol. 26, pp. 727–749, 1965.

7. S. Lefebvre, S. Richter, and R. DeCarlo, "Decentralized Variable Structure Control Design for a Two-Pendulum System," *IEEE Transactions on Automatic Control*, vol. AC-28, pp. 1112–1114, 1983.

8. D. D. Šiljak, *Decentralized Control of Complex Systems*, New York: Academic Press, 1991.

9. D. F. Wilkie and W. R. Perkins, "Generation of sensitivity functions for linear systems using low-order models," *IEEE Transactions on Automatic Control*, vol. AC-14, pp. 123–130, 1969.

Dynamic Modeling of Airborne Gas Turbine Engines

Ady Solomon

Pratt & Whitney Canada Inc., 1801 Courtney Park Drive, Mississauga, Ontario, Canada L5T 1J3
Email: ady.solomon@pwc.ca

Abstract. An airborne gas turbine engine with its accessories and installation frame constitutes the propulsion system of an airplane. In industrial practice, control system development for airborne gas turbine engines is part of an iterative converging process which involves control design and propulsion system dynamic testing. Competitive pressure to reduce development time and airframer demands for continuous improvements in propulsion performance have created the necessity to develop predictive reliable dynamic models which can simulate complex physical phenomena taking place inside a gas turbine engine. An example of such a dynamic model restricted to the core of an airborne gas turbine engine (gas generator) is presented in this paper. The physics based one-dimensional generic model can be used for propulsion system dynamic analysis tasks such as variable geometry optimization, air bleed extraction optimization, evaluation of component scaling effects, surge and rotating stall frequencies, determination of surge and rotating stall stability boundaries and for model based control design. The notion of surge domain is introduced and model predictions are given. The model can be partitioned to obtain an one-dimensional compressor dynamic model which can simulate onset and development of surge and demonstrate rotating stall boundaries. Simulation results which demonstrate the usefulness of the model are included.

1 Introduction

The stable operating range of a gas turbine based propulsion system is limited by aerodynamic flow instabilities such as surge and/or rotating stall which are initiated in the compression system. Recent combined research efforts of various engineering disciplines have provided more quantitative insight into the dynamics of these relatively high frequency phenomena [1]. In a classical sense the boundary of the compressor stable flow operation is called the surge line. The task of a control system designer is to provide a reliable algorithm to actuate propulsion system inputs as to insure its steady state and transient stable operation over the entire environmental envelope. Regardless of the approach used for the design of the control algorithm, i.e. classical or modern, the practicing control system designer needs a reliable dynamic model which can simulate propulsion system behavior when operating in unstable flow conditions occurring in the vicinity of the surge line. Dynamic models in turbomachinery are often obtained by combining compressor and turbine steady state characteristic maps with somehow empirically simplified gas dynamics differential equations (e.g. [2]). These models usually have a low frequency bandwidth and simulate faster phenomena such as surge or rotating stall in a qualitative way. In consequence the boundary of the stable flow operation domain is usually inferred from other facts and therefore the uncertainty in the position of the surge line may be quite large. When these models are used for control system design, approximate algorithms are obtained which may result in occasional instabilities in the propulsion system. In practice, control design limitations are corrected for by using tuning methods (e.g. [3]).

An one-dimensional (1D) control oriented physics based dynamic model which has a large enough frequency bandwidth to capture the onset and development of surge and demonstrate existence of rotating stall in a low speed axial compressor is described in [4–6]. The model is based on quasi-1D compressible flow equations.

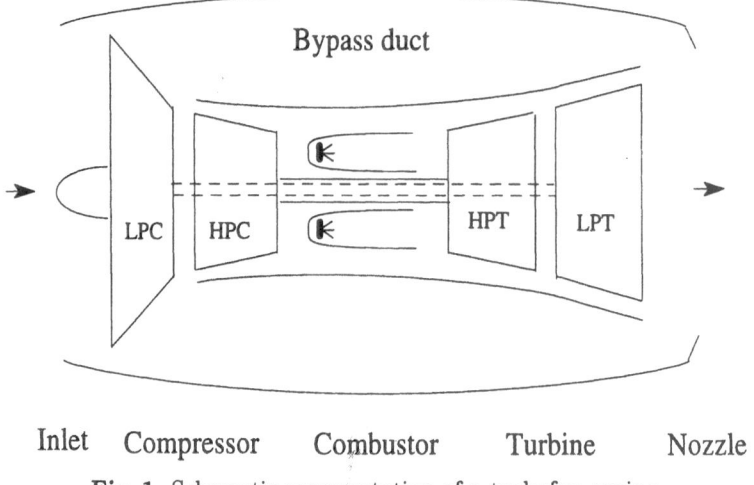

Fig. 1. Schematic representation of a turbofan engine.

In the present paper this approach has been extended so as to simulate the core of an airborne gas turbine engine (gas generator) from the Pratt & Whitney Canada PW300 series of turbofan engines. A schematic representation of a turbofan engine is given in Fig. 1 and that of a gas generator in Fig. 2. The full turbofan engine dynamic model can be obtained by adding to the core model that of the low pressure compressor (fan), the low pressure turbine, the bypass duct and the exit nozzle. In general terms, a gas generator turbine engine converts fuel energy into power which can be either mechanical shaft power or high speed thrust. It has three main sections: high pressure compressor, combustor and high pressure turbine. The PW300 gas generator is equipped with a multistage variable geometry compressor comprised of four axial and one centrifugal stage and a two stage turbine. Since this dynamic model simulates unstable flow conditions occurring in the vicinity of the surge

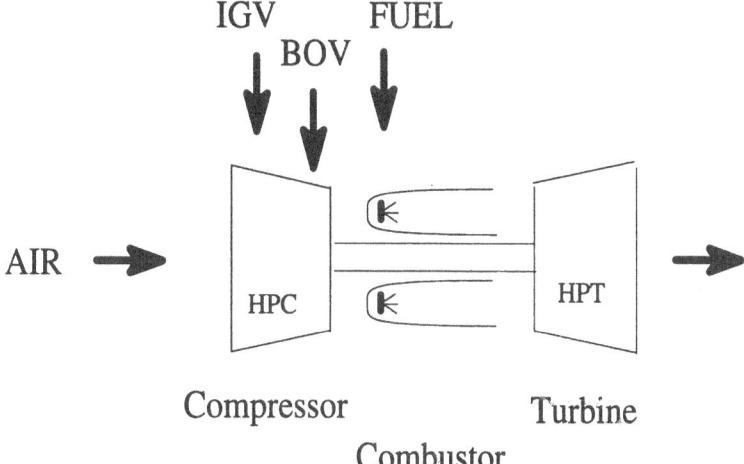

Fig. 2. Schematic representation of a gas generator engine.

line it can be used not only for control design but also for other system oriented design tasks such as: dynamic (transient) analysis, classical and modern nonlinear control design, variable geometry optimization, air bleed extraction optimization, off design point component optimization, acoustic and fuel flow noise effects.

2   Dynamic Model Development

The 1D control oriented dynamic model developed given in [4–6] for a low speed axial compressor has been used as the starting point in the development of a generic 1D dynamic model for a single shaft gas generator representative for the PW300 series engines. The gas generator dynamic model can be partitioned to obtain a stand alone compressor dynamic model for the first axial stage, first four axial stages or the entire high compressor.

The following notation will be used for nondimensional variables: A – flow area, L – length, M – Mach number, T – total temperature, e – specific internal energy, f – specific force, f_w – specific wall friction force, p – total pressure, Q – volumetric heat transfer rate, s – specific entropy, t – time, u – fluid velocity, x – axial coordinate, γ – heat capacity ratio.

The following notation will be used for dimensional variables: N – compressor shaft speed, R – gas constant, V_p - compressor plenum volume, ρ – density, c – sound velocity, c_p – heat capacity at constant pressure, c_v – heat capacity at constant volume, ω_H – Helmoltz resonator frequency.

The following subscripts notation will be used: a – actual physical quantity, i – input, o – output, s – static value.

The quasi-1D continuity, momentum and energy equations which describe the dynamics of the unsteady compressible, viscous flow of a calorically perfect gas are given by

$$\frac{\partial}{\partial t_a}(\rho_{sa}A_a) + \frac{\partial}{\partial x_a}(\rho_{sa}u_aA_a) = 0 \tag{1}$$

$$\frac{\partial}{\partial t_a}(\rho_{sa}u_aA_a) + \frac{\partial}{\partial x_a}\left[\left(\rho_{sa}u_a^2 + p_{sa}\right)A_a\right] = p_{sa}\frac{\partial}{\partial x_a}A_a + \rho_{sa}A_a(f_a + f_{wa}) \tag{2}$$

$$\frac{\partial}{\partial t_a}\left[\rho_{sa}\left(e_a + .5u_a^2\right)A_a\right] + \frac{\partial}{\partial x_a}\left[\rho_{sa}u_aA_a\left(e_a + \frac{p_{sa}}{\rho_{sa}} + 0.5u_a^2\right)\right]$$
$$= A_aQ_a + \rho_{sa}u_aA_af_a. \tag{3}$$

The equations for a calorically perfect gas are given by

$$p_{sa} = \rho_{sa}RT_{sa} \tag{4}$$
$$e_a = c_vT_{sa} \tag{5}$$
$$e_a + \frac{p_{sa}}{\rho_{sa}} = c_pT_{sa} \tag{6}$$

and the compressible flow functions for total pressure and temperature are given by

$$p_a = p_{sa}\left(1 + \frac{\gamma - 1}{2}M^2\right)^{\frac{\gamma}{\gamma-1}} \tag{7}$$

$$T_a = T_{sa}\left(1 + \frac{\gamma - 1}{2}M^2\right) \tag{8}$$

where

$$M = \frac{u_a}{\sqrt{\gamma RT_{sa}}}, \qquad \gamma = \frac{c_p}{c_v}.$$

Let $p_{ref}, T_{ref}, t_{ref}$ be total pressure, temperature and time constant reference values respectively. Then the non-dimensional variables p, T and t are given by

$$p = \ln\left(\frac{p_a}{p_{ref}}\right), \quad T = \ln\left(\frac{T_a}{T_{ref}}\right), \quad t = \frac{t_a}{t_{ref}}. \tag{9}$$

The entropy is defined using Gibb's equation applied to a calorically perfect gas

$$s_a = c_p\ln\left(\frac{T_a}{T_{ref}}\right) - R\ln\left(\frac{p_a}{p_{ref}}\right); \tag{10}$$

let s be defined as

$$s = \frac{s_a}{c_p}.$$

Equations (1)-(10) hold true for any section of the gas generator gas path. For simplicity of presentation assume without loss of generality that (1)-(10) have been written for a compressor stage (rotor, stator). Let the steady state compressor maps be expressed as a function of M_o, N as follows

$$\pi_M = \pi_M(M_o, N), \quad \pi_p = \pi_p(M_o, N), \quad \pi_s = \pi_s(M_o, N). \tag{11}$$

Then it can be shown analytically (e.g. [4,5]) that the dynamic variation of the total pressure, temperature and Mach number at the input and output of the stage is given by the following system of nonlinear differential equations

$$\varepsilon(s_i, \ p_i, \ L)\frac{d}{dt}\begin{bmatrix} M_i \\ p_o \\ s_o \end{bmatrix}$$

$$= \Xi(M_o, \gamma)\left\{ \begin{bmatrix} M_o - M_i \\ p_o - p_i \\ s_o - s_i \end{bmatrix} - \begin{bmatrix} \pi_M \\ \pi_p \\ \pi_s \end{bmatrix} \right\} + \varepsilon(s_i, \ p_i, \ L)\begin{bmatrix} 0 \\ -1 \\ \frac{\gamma-1}{\gamma} \end{bmatrix}\frac{d}{dt}A_o \tag{12}$$

where the entries of the matrix $\Xi(M, \gamma)$ are the influence coefficients which are known functions defined in [6],

$$\varepsilon(s_i, \ p_i, \ L) = Le^{-0.5(s_i + \frac{\gamma-1}{\gamma}p_i)}$$

is a scalar and L is the non-dimensional length of the stage [4].

The system of equations (12) is generic in the sense that any section of the gas path can be similarly modeled. The set of all models for each gas path section constitutes the dynamic model of turbomachinery hardware considered.

3   Gas Generator Dynamic Model Description

The PW300 gas generator dynamic model is partitioned in dynamic stages and each stage has a generic representation as given by (12). A dynamic stage is defined as the spatial domain in the gas generator where the nonlinear partial differential equations of mass continuity, momentum and energy are solved. The compressor is comprised of two variable geometry axial stage, two fixed geometry axial stages and one centrifugal stage and the turbine has two stages. In the dynamic model each compressor stage is also a dynamic stage while for the turbine each blade row is a dynamic stage. The combustor is modeled as one dynamic stage. In the present model all air bleed extractions as well as the cooling air bleeds are modeled to satisfy the mass continuity equation only. It should be noted that the term $\frac{dA_o}{dt}$ in (12) is nonzero for the first variable geometry stage of the compressor.

The 1D gas generator model uses stage by stage compressor and turbine steady state characteristic maps generated by mean-line programs and the actual geometrical dimensions of the hardware. At each dynamic stage inlet and outlet the model calculates steady state and transient values of total pressure, temperature and Mach number. Other physical quantities of interest can be obtained as well, e.g. mass air flow, corrected mass air flow, shaft speed, turbine power, static pressures and temperatures. From a systems point of view the input to the gas generator model is fuel flow, inlet guide vanes position, amount of bleed extraction and exit throttle valve area.

In addition to control system design the model can be used for optimization of variable geometry to enhance surge margin, optimization of variable geometry and bleed schedule to achieve engine acceleration in the required time (e.g. 5 sec), effects of components scaling, acoustic and fuel flow noise effects, feasibility studies of various control design approaches.

4 Dynamic Model Surge Domain Prediction

Aerodynamic instabilities may cause a compressor to surge or enter rotating stall. Surge is an axial oscillation in the compressor characterized by a complete breakdown of the air flow. Following a surge, a compressor may or may not recover its stable operation. Rotating stall is a circumferential pressure oscillation which is initiated in a compressor blade row and is characterized by a partial obstruction of the axial air flow through the compressor. Depending on compressor design, a rotating stall type instability may die out, may cause the compressor to surge or may cause the compressor to operate quasi-stable at an undesired operating point. In practice, the boundary of the stable operation of the compressor is called surge line and is usually plotted on the steady state compressor map. A typical steady state compressor map consists of a family of constant speed curves (pressure ratio versus mass air flow). Pressure ratio is typically the ratio of total pressure at compressor outlet and inlet.

Theoretical and experimental insight gained in the last few years into the fluid dynamic phenomena occurring in the vicinity of the surge line suggests that actually there exists a surge domain. A surge domain is defined as the upper limit of the stable compressor operation region: the lower bound of the surge domain is the steady state surge line and its upper bound is the transient surge line. The steady state surge line is a boundary on the compressor map such that if the compressor operating point is located above it then the compressor may surge, may enter rotating stall or may remain stable. The transient surge line is a boundary on the compressor map such that if the compressor operating point is located above it then the compressor will always surge. The classical surge line definition corresponds to the present steady state surge line definition. Fig. 3 shows a schematic representation of the surge domain for a compressor map in the coordinates pressure ratio (PR) versus mass air flow (W) at two constant shaft speeds.

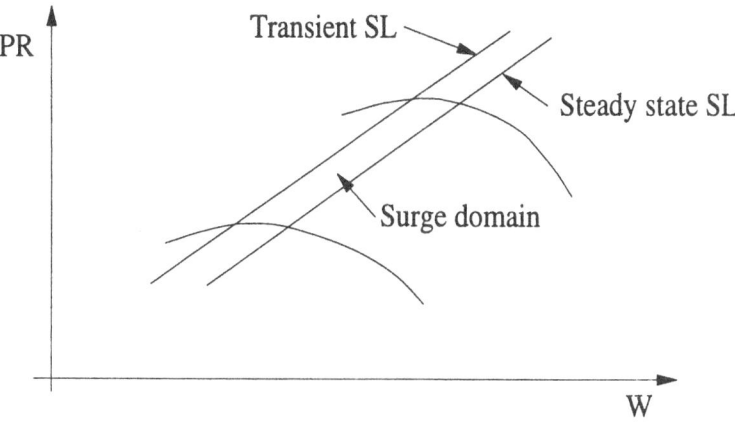

Fig. 3. Surge domain definition.

In practice it is difficult to measure the transient surge line of a compressor. Therefore an important aspect of the quality of the dynamic model is its ability to predict the position of the steady state surge line and the surge domain for a given compressor map. The compressor dynamic model has been used to simulate surges and to obtain the surge domain. For a given constant air bleed extraction, constant shaft speed and constant position of the variable geometry the exit throttle area is suddenly closed (step change) from the running line set point to a given set point at which a very small amplitude pressure oscillation will appear at the compressor exit. This is a point on the steady state surge line. For the same bleed extraction, shaft speed and variable geometry position the exit throttle area is suddenly closed (step change) from the running line set point to a given set point closer than the previous one until a surge will be simulated. This is a point on the transient surge line. The above procedure is repeated for a set of shaft speeds and variable geometry positions until the surge line and the surge domain is satisfactorily mapped.

Let the nondimensional pressure and flow coefficients be given by ([10,11])

$$\Psi_s = (p_{so} - p_i)/\left(\frac{1}{2}\rho U^2\right) \text{ and } \Phi = C_x/U \tag{13}$$

respectively. Then it has been shown in [7] that for a single stage compression system, the steady state surge line is the locus of all the points which satisfy the relationship $\frac{d\Psi_s}{d\Phi} = 0$.

The schematic diagram of the dynamic model of the first compression stage is given in Fig. 4. The steady state surge line has been determined from the dynamic model using step changes of the exit throttle set points. The surge line position predicted by the 1D dynamic model has been compared with that obtained analytically by using (13). For the case of inlet guide vanes positioned

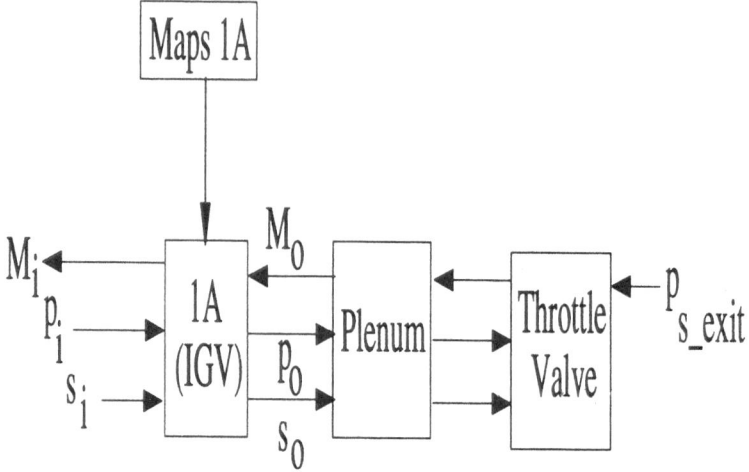

Fig. 4. Schematic diagram of the first stage compressor dynamic model.

at zero ($IGV = 0$), the results given in terms of pressure ratio versus mass air flow (W) are shown in Fig. 5. The two results are very close each to the other.

Simulations of the 1D dynamic model of the high pressure compressor show that for a given set of shaft speeds and variable geometry with no bleed extraction the steady state surge line predicted by dynamic model is quite close to that found experimentally on PW300 gas generator engine. The simulated steady state and transient surge line are shown in Fig. 6 in terms of compressor pressure ratio versus shaft speed for standard total pressure and temperature at compressor inlet of 14.696 psi and 59° F respectively. The surge domain is the area in between the two lines. The experimentally determined surge line is almost entirely within the surge domain defined by the dynamic model (Fig. 7).

5   Surge and Rotating Stall

The compressor dynamic model can be used to simulate surge propagation through the compressor and to demonstrate the existence of a rotating stall. In general terms, surge is a longitudinal oscillation in the compressor which causes a sudden loss of pressure at compressor exit. Rotating stall is a circumferential oscillation usually occurring at one compressor stage which causes an unstable operation of the compressor.

Figure 8 gives an example of surge instability and its propagation along the compressor at 91.8% shaft speed, standard inlet conditions and a given variable geometry setting. The simulated total pressure at the exit of each compressor stage is given as a function of time expressed in rotor revolutions. Surge inception, surge propagation and surge timing are the primary parameters of interest while surge amplitude is of secondary interest.

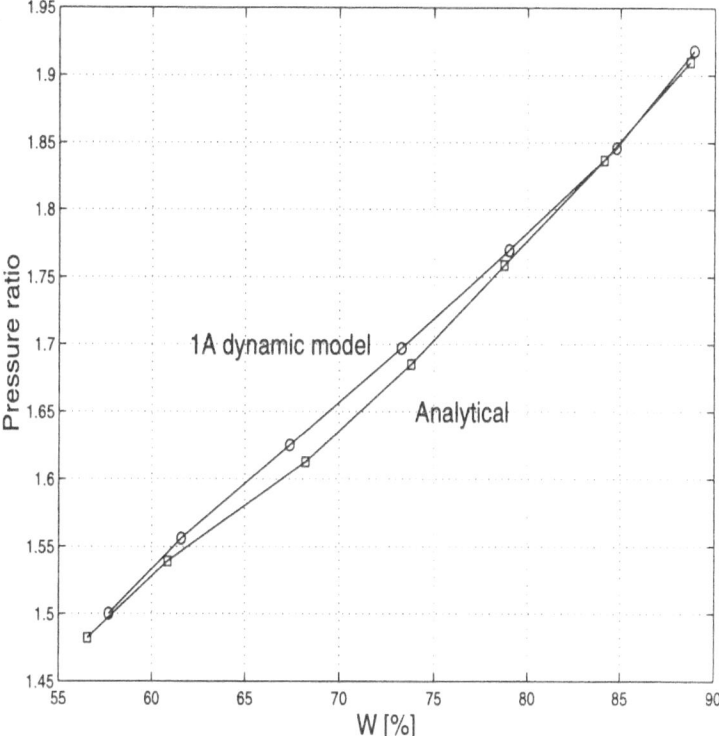

Fig. 5. Surge line prediction for stage 1A, IGV=0.

Referring to Fig. 8, the throttle valve is stepped towards closing (smaller throttle area) at time 83 rotor revolutions. The simulation indicates that it takes about 25 rotor revolutions to fully recover from a single surge. From the compressor design it is known that at this speed the compressor centrifugal stage is most likely to surge first. This fact is evident from Fig. 8 where the surge event which starts in the centrifugal compressor at time 88 rotor revolutions (bottom trace) occurs before the pressure rise in the axial stages is completed.

Rotating stall is a nonaxisymmetric phenomena best simulated by a two dimensional model (e.g. [8]). An 1D dynamic model such as that described in this paper can demonstrate the existence of a rotating stall in a compressor. Fig. 9 shows the simulated total pressure at the exit of each compressor stage at 74.3% shaft speed, all bleeds closed and standard inlet conditions. Rotating stall events occur at time 78, 88 and 100 rotor revolutions. The propagation of rotating stall along the compressor x-axis is best illustrated at time 100 rotor revolutions. Stage 3A is in rotating stall which propagates forwards and backwards as predicted by the quantitative analysis developed in [9]. Fig. 9 indicates that rotating stall at stages 2A and 4A has almost the same shape and

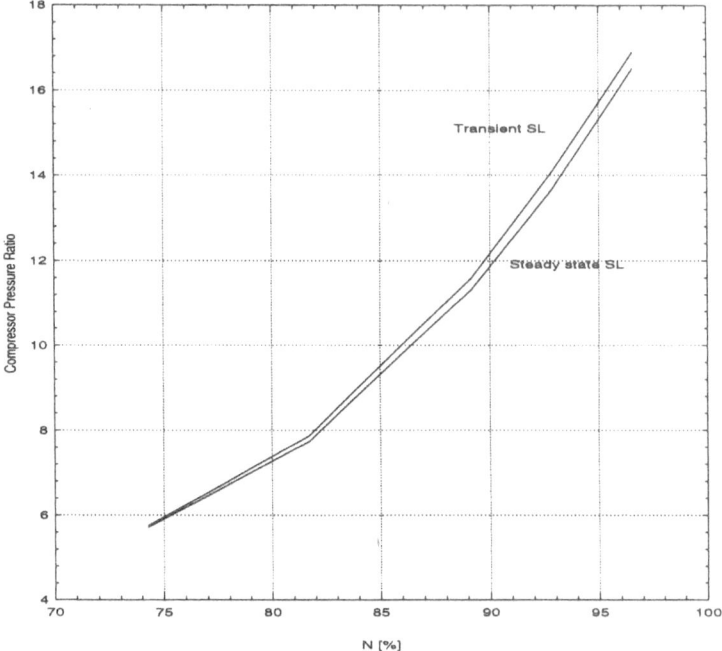

Fig. 6. Compressor dynamic model transient and steady state surge line.

amplitude as that in stage 3A. At stage 1A and 1C rotating stall is attenuated as expected. Hence it is most likely that rotating stall is initiated in stage 3A at this speed.

In terms of computation time, it takes about 49 seconds on an HP 9000/715 workstation to simulate 700 msec of real time compressor operation.

6   Surge Frequency at Constant Speed

The Helmoltz frequency for given compressor is defined in [10,11] as

$$\omega_H = c\sqrt{\frac{A_c}{V_p L_c}} \tag{14}$$

where A_c and L_c are the area and the length respectively of an equivalent compressor duct. For a given compressor and inlet conditions, at constant compressor speed the Helmoltz frequency (eqn. 14) can be written as

$$\omega_H = K V_p^{-\frac{1}{2}} \tag{15}$$

where K is a constant

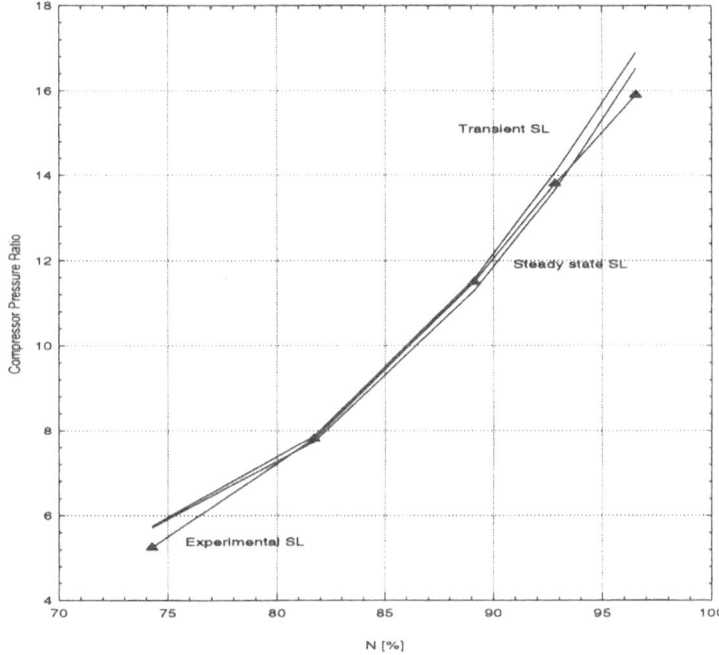

Fig. 7. Experimental surge line and compressor dynamic model surge lines.

The use of (14) to calculate the surge frequency is not straightforward since for a multistage compressor the area and length of an equivalent compressor are not readily available. Therefore (15) has been used instead to compare the ratio of surge frequencies for various plenum volumes.

The compressor dynamic model has been used to simulate a surge at N=91.8% for two plenum volumes: nominal volume equal to the actual compressor plenum volume and another smaller volume (Fig. 10). The ratio of the two volumes is 7/4. From Fig. 10 the surge frequency for the nominal volume is 16.67 Hz and that for the smaller volume is 25 Hz which gives a ratio of 0.668. The predicted surge frequency ratio from eqn.(14) is 0.756 which yields a 14% difference between the ratios. Surge frequencies recorded during full engine tests are close to those predicted by the dynamic model. Hence via the dynamic model a link has been established among surge frequency theoretical prediction, model prediction and experimental data.

7 Air Bleed Extraction Optimization

Air bleed extraction from a compressor results in a lower operating line. In consequence the surge margin will increase and the compressor operation will become more stable. In practice it is desired to be able to find by design the

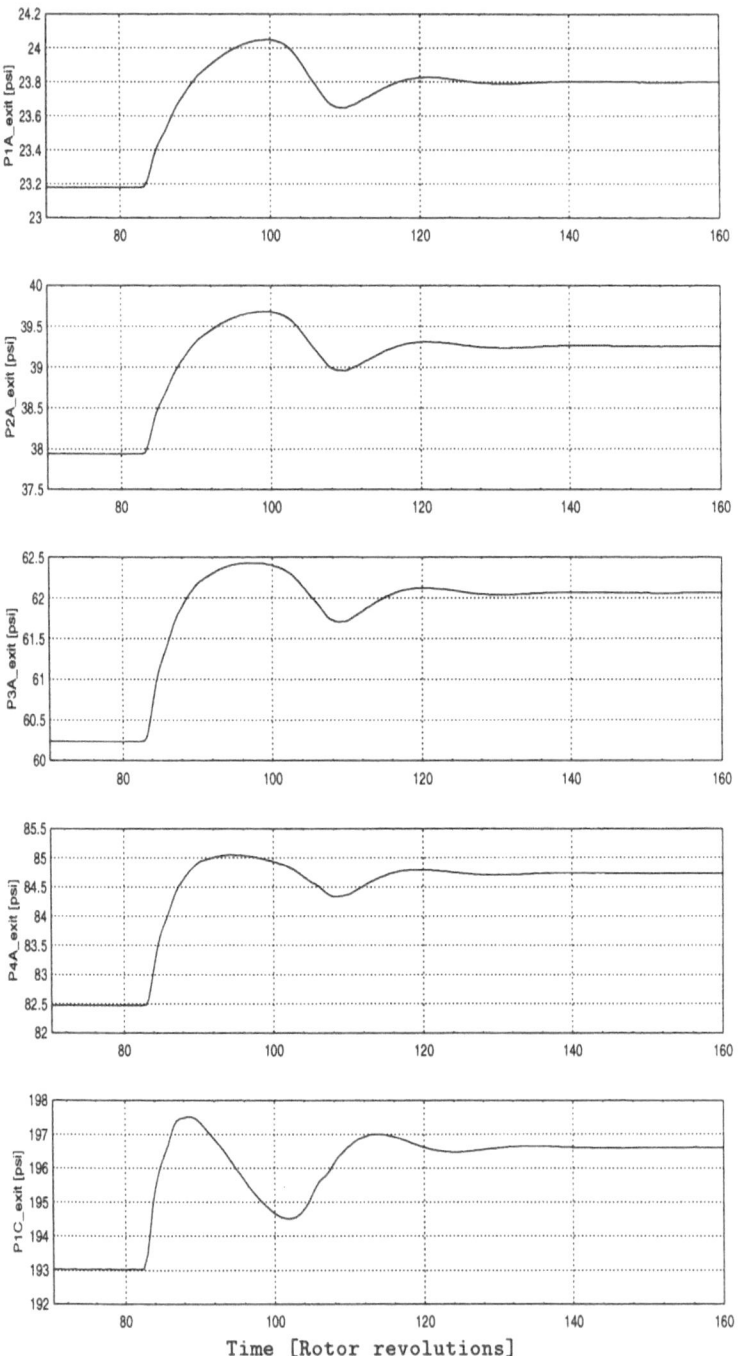

Fig. 8. Surge propagation along compressor stages at 91.8% shaft speed.

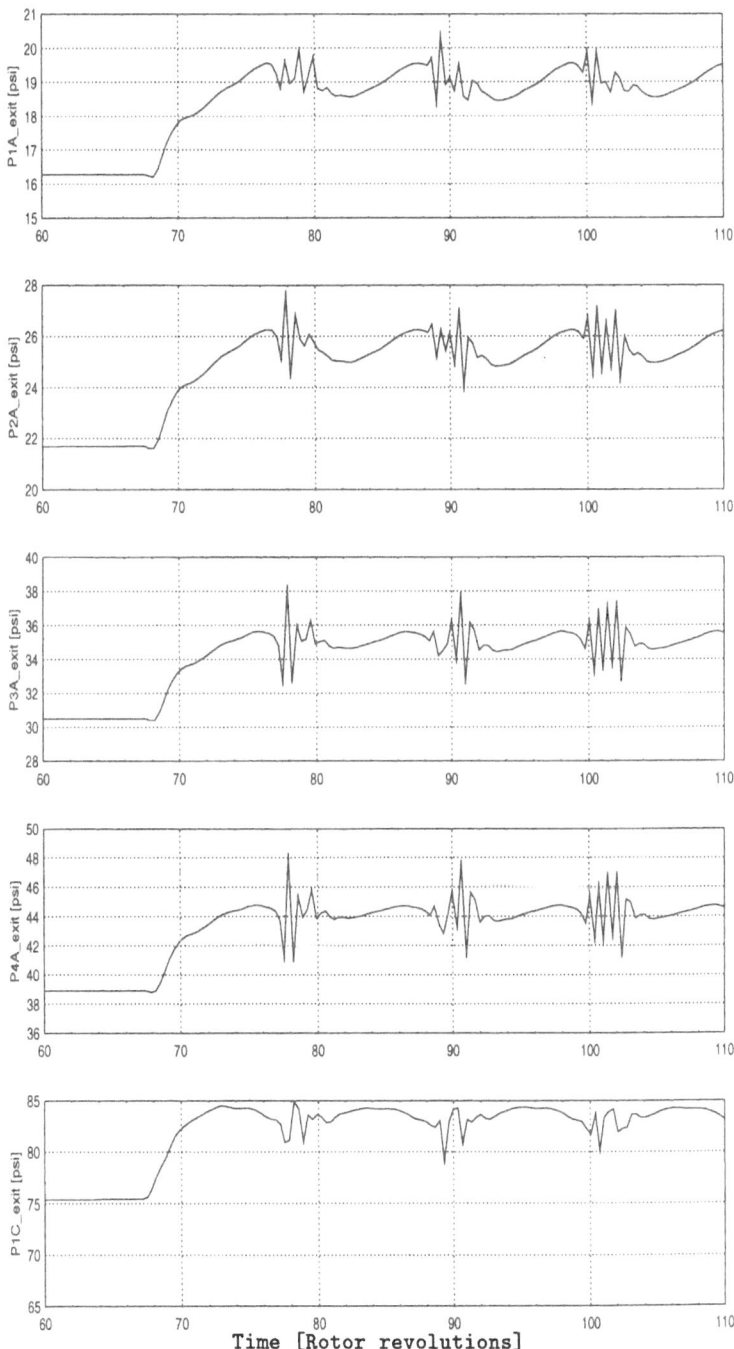

Fig. 9. Simulation results indicating existence of rotating stall at 74.8% shaft speed.

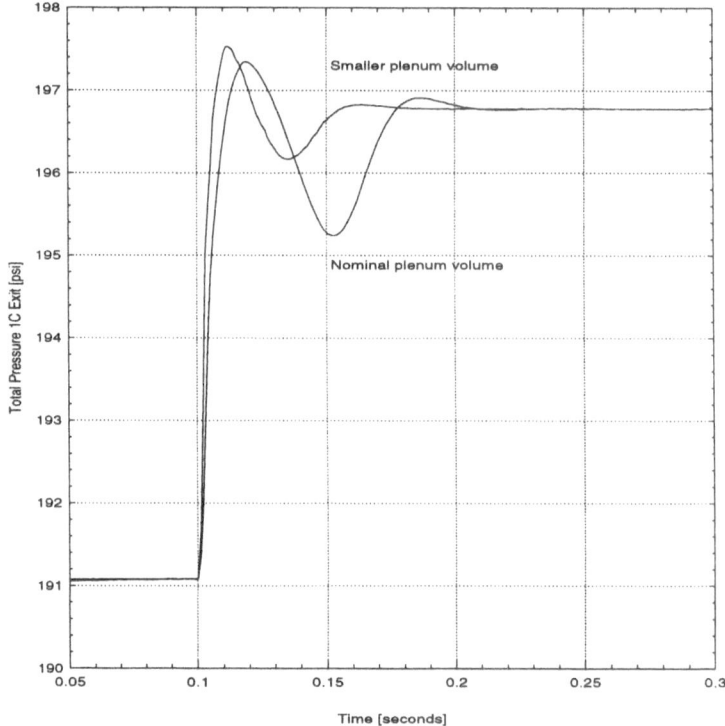

Fig. 10. Simulated surge event for nominal and smaller plenum volume.

optimum amount of air bleed necessary to ensure a stable operation of the compressor during transients.

An example of steady state air bleed extraction optimization from the exit of stage 4A at N=81.1%, given variable geometry schedule and standard inlet conditions is given in Fig. 11. At time 0.1 sec the throttle area is decreased (step change) and the compressor model exhibits multiple surges (Fig. 11(a)) until time 0.6 sec when the throttle valve area is reset to its initial value and surges are cleared. Fig. 11(b) shows that the amplitude of surges decreases when 2% of air flow is extracted while the surges are completely cleared with 4% of air bleed extraction (Fig. 11(c)).

8   Gas Generator Dynamic Simulation

The exit nozzle of the gas generator dynamic model has been calibrated at steady state such that for the same inputs applied to the experimental gas generator the model will accurately simulate the actual measured performance. Simulations have been performed to validate the performance of the dynamic model. An example of the response of the model and that of the experimental

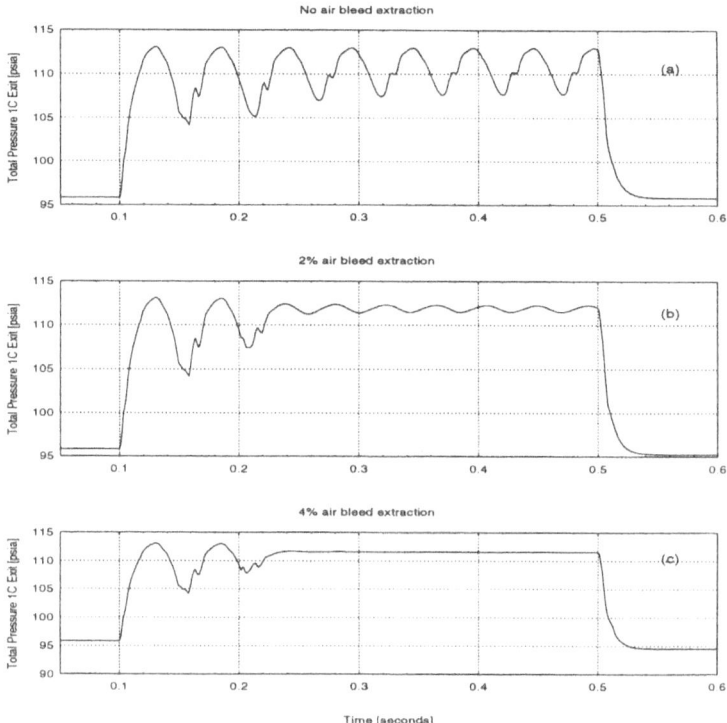

Fig. 11. Simulated surge event for different bleed extractions.

gas generator to a WF (fuel flow) impulse input for the case of constant IGV (inlet guide vanes) angle and BOV (bleed off valve) closed at N=24 krpm is given in Fig. 12. The recorded WF impulse input to the experimental gas generator (Fig. 12(a)) has been applied off line in non-real time to the gas generator model. The variation in time of the gas generator shaft speed and the pressure at HPC exit in response to the WF impulse are shown in Fig. 12(b) and (c).

The accuracy with which the model reproduces the experimental data implies that the dynamic model has a sufficiently large frequency bandwidth and it can simulate other dynamic aspects related to the gas generator or full engine such as variable geometry optimization, noise effects, components scaling effects, analysis of design trade offs.

9   Conclusions

A generic physics based one-dimensional nonlinear dynamic model of a gas generator has been developed which can be easily partitioned to obtain the dynamic model of the entire compression system or of any given number of

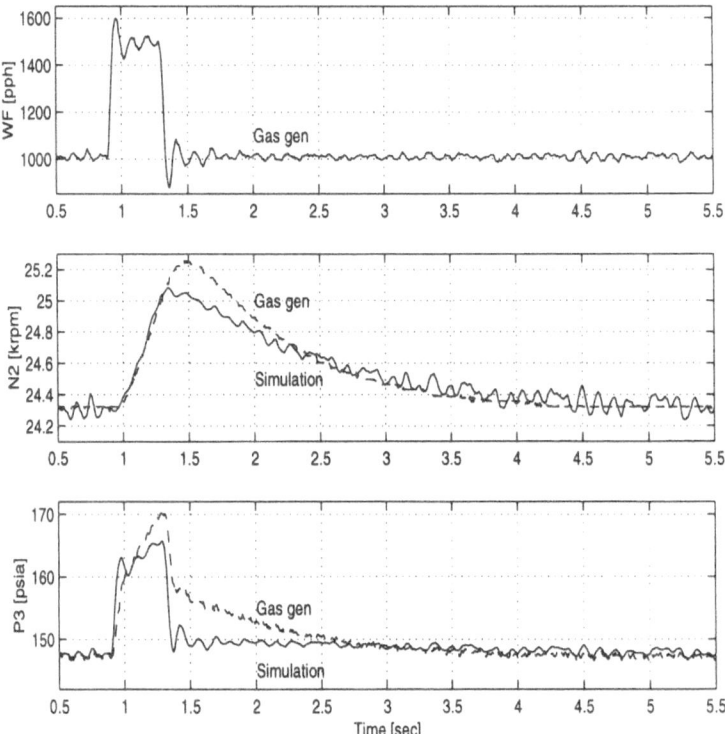

Fig. 12. Gas generator test data and simulation results at N=24 krpm, WF impulse, IGV frozen, BOV closed.

stages. By adding the model of the low pressure compressor (fan) and low pressure turbine, the dynamic model of a turbofan engine can be obtained. The compression system dynamic model can predict the surge domain, it can simulate surges and rotating stalls and predict surge frequency. In addition to control design, the gas generator or full engine dynamic model can be used for other design tasks such as variable geometry optimization, air bleed extraction optimization, evaluation of component scaling effects, acoustic and fuel noise effects, and off design point component optimization.

Acknowledgment

The author would like to thank Pratt and Whitney Canada for the permission to publish this paper. Also would like to thank my colleagues from Pratt and Whitney Canada who indirectly contributed through useful discussions to the successful completion of this model.

References

1. J. P. Longley, "A review of non-steady flow models for compressor stability," *International Gas Turbine and Aeroengine Congress and Exposition*, Cincinnati, Ohio, May 24–27, 1993.

2. D. Garrard, "ATEC: the aerodynamic turbine engine code for the analysis of transient and dynamic gas turbine engine system operations, Part 1: model development," *ASME International Gas Turbine Institute's Turbo Expo'96*, NEC, Birmingham, UK, June 10–13, 1996.

3. A. Solomon, "Full authority digital electronic control of Pratt and Whitney 305 turbofan engine," *SAE paper 911018, SAE General, Corporate, & Regional Aviation Meeting & Exposition*, April 9–11, 1991.

4. O. O. Badmus, K. M. Eveker, C. N. Nett, "Control-oriented high-frequency turbomachinery modeling: general 1D model development," *ASME paper 93-GT-385, International Gas Turbine and Aeroengine Congress and Exposition*, Cincinnati, Ohio, May 24–27, 1993.

5. O. O. Badmus, S. Chowdhury, K. M. Eveker, C. N. Nett, "Control-oriented high-frequency turbomachinery modeling: single-stage compression system 1D model," *ASME paper 93-GT-18, International Gas Turbine and Aeroengine Congress and Exposition*, Cincinnati, Ohio, May 24–27, 1993.

6. O. O. Badmus, K. M. Eveker, C. N. Nett, "Control-oriented high-frequency turbomachinery modeling, Part I: theoretical foundations," *Proceedings of the AIAA Joint Propulsion Conference*, AIAA paper 92-3314, 1992.

7. A. H. Stenning, "Rotating stall and surge," *ASME Journal of Fluids Eng.*, 102, pp. 14–21, 1980.

8. M. R. Feulner, G. J. Hendricks, J. D. Paduano. "Modeling for control of rotating stall in high speed multi-stage axial compressors," *International Gas Turbine and Aeroengine Congress and Exposition*, The Hague, Netherlands, June 13–16, 1994.

9. V. H. Garnier, A. H. Epstein, E. M. Greitzer, "Rotating waves as a stall inception indication in axial compressors," *Transactions of the ASME, Journal of Turbomachinery*, vol. 113, pp. 290–302, 1991.

10. E. M. Greitzer, "Surge and rotating stall in axial flow compressors, Part I: theoretical compression system model," *Transactions of the ASME, Journal of Engineering for Power*, pp. 190–198, 1976.

11. E. M. Greitzer, "Surge and rotating stall in axial flow compressors, Part II: experimental results and comparison with theory," *Transactions of ASME, Journal of Engineering for Power*, pp. 199–217, 1976.

Properties of Linear Discrete-Time System in Terms of Its Singular Value Decomposition

Shih-Ho Wang

Department of Electrical and Computer Engineering, University of California,
Davis, CA 95616, USA
Email: wang@ece.ucdavis.edu

Abstract. In this paper, we study the relationship between the non-minimum phase property of a linear discrete-time system and the output reachability of the system. We apply the singular value decomposition to the input-output matrix of a linear discrete-time system. This approach provides a geometrical interpretation of the reachable subspace of the system output. For the time-invariant case, we also prove certain symmetry properties between the singular vectors of the system input-output matrix.

1 Introduction

In the control of linear time-invariant system, it is well-known that a system transfer function with unstable zeros (non-minimum phase system) exhibits certain undesirable properties and is difficult to control in general. It is unclear how to characterize a non-minimum phase system in the time-domain. Furthermore, it is not clear how to generalize the concept of non-minimum phase system to the time-varying case.

In this paper, we consider the case of linear time-varying system over a finite time horizon. The input-output relationship of the system is represented by a constant matrix. Then one computes the singular value decomposition of the input-output matrix. Based on the singular values and the condition number of the matrix, it clearly shows how difficult it is to invert such a matrix. This coincides with the general problem of controlling a system, i.e., finding the appropriate input so that the system produces the desired output. In the case of linear time-invariant system, we show the direct correspondence between the non-minimum phase zeros and the singular values of the input-output matrix. Hence, the concept of non-minimum phase system can be generalized to the time-varying case by using the singular value decomposition. In this paper, we also show certain symmetry properties between the singular vectors of the system input-output matrix.

2 Singular Value Decomposition

Consider a linear time-varying system with m inputs and p outputs,

$$x(i+1) = A(i)x(i) + B(i)r(i)$$
$$y(i) = C(i)x(i) + D(i)r(i)$$

where $A(i)$, $B(i)$, $C(i)$ and $D(i)$ are real matrices of proper size, $r(i) \in \mathbb{R}^m$ is the input vector, $x(i) \in \mathbb{R}^n$ is the state vector, $y(i) \in \mathbb{R}^p$ is the output vector, and $i = 0, 1, \ldots$.

Assume the initial state $x(0) = 0$. Then for any positive integer N, we can form the following input-output relationship:

$$
\begin{bmatrix} y(0) \\ y(1) \\ \vdots \\ y(N-2) \\ y(N-1) \end{bmatrix} = \begin{bmatrix} h_{00} & 0 & \cdots & 0 & 0 \\ h_{10} & h_{11} & \ddots & 0 & 0 \\ \vdots & & \ddots & \ddots & \vdots \\ h_{(N-2)0} & h_{(N-2)1} & \ddots & h_{(N-2)(N-2)} & 0 \\ h_{(N-1)0} & h_{(N-1)1} & \cdots & h_{(N-1)(N-2)} & h_{(N-1)(N-1)} \end{bmatrix} \begin{bmatrix} r(0) \\ r(1) \\ \vdots \\ r(N-2) \\ r(N-1) \end{bmatrix}
$$

where

$$
h_{ij} = \begin{cases} D(i) & \text{if } i = j, \\ C(i)A(i-1)\cdots A(j+1)B(j) & \text{if } i > j. \end{cases}
$$

The above equation is written as follows:

$$
y = Hr
$$

where

$$
y = \left[y^T(0)\ y^T(1)\ \cdots\ y^T(N-2)\ y^T(N-1) \right]^T \in \mathbb{R}^{pN}
$$

and

$$
r = \left[r^T(0)\ r^T(1)\ \cdots\ r^T(N-2)\ r^T(N-1) \right]^T \in \mathbb{R}^{mN}
$$

are the augmented input and output vectors, and $H \in \mathbb{R}^{pN \times mN}$ is the input-output matrix. If we apply singular value decomposition to H, then

$$
H = U\Sigma V^T
$$

where U and V are both orthonormal matrices, and $\Sigma = \text{diag}(\sigma_1, \sigma_2, \ldots, \sigma_{sN})$ with non-negative singular values on the diagonal, where $s = \min\{p, m\}$.

Let u_i and v_i, $i = 1, \ldots, sN$, be the columns of U and V, respectively. Then H can be written as

$$
H = \sigma_1 u_1 v_1^T + \cdots + \sigma_{sN} u_{sN} v_{sN}^T.
$$

Because U and V are orthonormal matrices, the set of vectors u_i form an orthonormal basis for the space of the output vector y, and the set of vectors v_i form an orthonormal basis for the space of the input vectors r.

From the system input-output point of view, if the input $r = v_i$, then the output $y = \sigma_i u_i$. Hence, if some of the singular values σ_i are close or equal to zero, then the input required to produce the output must have very large magnitude, or be infinite.

For a given input r, we can decompose it onto the set of basis vectors v_i, i.e.,

$$r = \sum_i \langle r, v_i \rangle v_i,$$

then the output y is

$$y = Hr = \sum_i \langle r, v_i \rangle H v_i = \sum_i \langle r, v_i \rangle \sigma_i u_i$$

where $\langle \cdot, \cdot \rangle$ is the inner product operation.

Conversely, for a given output y, we can decompose it onto the set of basis vectors u_i, i.e.,

$$y = \sum_i \langle y, u_i \rangle u_i.$$

The corresponding input r is

$$r = \sum_i \frac{\langle y, u_i \rangle}{\sigma_i} v_i$$

assuming that all $\sigma_i > 0$.

Therefore, based on the set of singular vectors $\{u_1, \ldots, u_{pN}\}$, we have a clear picture of which "direction" in the output function space is achievable. For an illustrative example, please refer to [3].

3   Non-Minimum Phase Zero and Output Reachability

For a given system, in order to define its output reachability, one has to be careful in defining its input and output spaces. For instance, consider a system consisting of a single delay element,

$$x(i+1) = r(i)$$
$$y(i) = x(i).$$

In this case, $A(i) = D(i) = 0$, and $B(i) = C(i) = 1$. Hence the system input-output is related by

$$
\begin{bmatrix} y(0) \\ y(1) \\ \vdots \\ y(N-1) \end{bmatrix}
=
\begin{bmatrix} 0 & \cdots & 0 & 0 \\ 1 & \cdots & 0 & 0 \\ \vdots & \ddots & \vdots & \vdots \\ 0 & \cdots & 1 & 0 \end{bmatrix}
\begin{bmatrix} r(0) \\ r(1) \\ \vdots \\ r(N-1) \end{bmatrix}.
$$

If we choose the output space as $\left[y(0)\ y(1)\ \cdots\ y(N-1)\right]$, then the input-output matrix is singular, and the system is not output reachable. Instead, if we choose the output space as $\left[y(1)\ y(2)\ \cdots\ y(N-1)\right]$, then the input-output matrix is of full rank and the system is output reachable, i.e., one can easily generate any output sequence $y(1), y(2), \ldots, y(N-1)$ by applying appropriate input sequence $r(0), r(1), \ldots, r(N-1)$.

According to the usual definition, the above system is a non-minimum phase system, since it has a zero at the infinity. However, depending on the selection of the output space, one may be able to generate any desired output without any difficulty. Therefore, analiying a system in the time domain allows us to investigate it in greater details.

4 Symmetry Between the Singular Vectors of the System Input-output Matrix

For linear time-invariant single input and single output system, the input-output matrix has the following special form:

$$
H = \begin{bmatrix}
h_0 & 0 & \cdots & 0 & 0 \\
h_1 & h_0 & \ddots & 0 & 0 \\
\vdots & \ddots & \ddots & \ddots & \vdots \\
h_{N-2} & h_{N-3} & \ddots & h_0 & 0 \\
h_{N-1} & h_{N-2} & \cdots & h_1 & h_0
\end{bmatrix}.
$$

In this section, we will prove certain symmetry property between the left and right singular vectors of H. Let

$$
\bar{I}_N = \begin{bmatrix}
0 & \cdots & 1 \\
\vdots & \ddots & \vdots \\
1 & \cdots & 0
\end{bmatrix}
$$

be a $N \times N$ matrix with 1's on the anti-diagonal and 0's elsewhere.

Lemma 1. *With H and $\bar{I}_N$ as defined above, then*

(a) $\bar{I}_N \bar{I}_N = I_N$ *(the $N \times N$ identity matrix)*,
(b) $\bar{I}_N H \bar{I}_N = H^T$,
(c) $\bar{I}_N H H^T \bar{I}_N = H^T H$.

Proof. (a) and (b) are obvious. For (c),

$$
\bar{I}_N H H^T \bar{I}_N = \bar{I}_N H (\bar{I}_N \bar{I}_N) H^T \bar{I}_N = (\bar{I}_N H \bar{I}_N)(\bar{I}_N H^T \bar{I}_N) = H^T H.
$$

$\square$

The following theorem proves the symmetry property between the left and right singular vectors of H.

Theorem 1. *Let H be the input-output matrix of a single input and single output linear time-invariant system. Its singular value decomposition is*

$$H = U \Sigma V^T. \tag{1}$$

Let u_i and v_i be the i-th column of U and V, respectively. Assume that the singular values of H are distinct. Then

$$v_i = \pm \bar{I}_N u_i, \; (i = 1, \dots, N).$$

Proof. From (1)

$$H H^T = U \Sigma V^T V \Sigma U^T = U \Sigma^2 U^T \tag{2}$$

and

$$H^T H = V \Sigma U^T U \Sigma V^T = V \Sigma^2 V^T. \tag{3}$$

From (2) and (3)

$$(H H^T - \sigma_i^2 I_N) u_i = 0 \tag{4}$$
$$(H^T H - \sigma_i^2 I_N) v_i = 0 \tag{5}$$

where σ_i is the i-th singular value of H. Using Lemma 1 and (4),

$$\bar{I}_N (H^T H - \sigma_i^2 I_N) \bar{I}_N u_i = 0. \tag{6}$$

This implies

$$(H^T H - \sigma_i^2 I_N)(\bar{I}_N u_i) = 0. \tag{7}$$

Since the singular values σ_i are distinct, the dimension of the null space of the following matrix

$$H^T H - \sigma_i^2 I_N$$

is 1. From (5) and (7),

$$v_i = c \bar{I}_N u_i$$

for some non-zero constant c. Since both vectors are normalized to 1, hence $c = \pm 1$. □

Remark 1. The above theorem states that the left and right singular vectors v_i and u_i satisfies the following relationship:

$$\begin{bmatrix} v_{i1} & v_{i2} & \cdots & v_{iN} \end{bmatrix} = \pm \begin{bmatrix} u_{iN} & u_{i(N-1)} & \cdots & u_{i1} \end{bmatrix}.$$

Again, for an illustrative example, please refer to [3].

5   Conclusions

In this paper, we apply singular value decomposition to the system input out-put matrix for linear discrete-time systems. The set of the singular values corresponds to the system gains. Therefore, one can define the reachable sub-space of the system output as spanned by the set of singular vectors with singular values greater than some constant ε. We have shown certain symme-try property among the singular vectors for linear time invariant systems. Our approach should prove to be most useful in analyzing the transient response of a system, as the Fourier analysis approach is useful for the steady-state analysis of a system.

References

1. G. Golub and C. Van Loan, *Matrix Computations*, the Johns Hopkins University Press, 1989.
2. G. Golub and W. Kahan, "Calculating the singular values and pseudo-inverse of a matrix," *SIAM J. Num. Anal.*, vol. 2, pp. 205–224, 1965.
3. S. H. Wang and T. F. Lee, "A new interpretation of non-minimum phase system with extension to time-varying case," *Electronics Letters*, vol. 32, pp. 270–272, 1996.